Value Creation
Edited by

McKinsey & Company

F. Budde, G. A. Farha,
H. Frankemölle,
D. F. Hoffmeister, K. Krämer

Value Creation

Strategies for the Chemical Industry

Edited by

McKinsey & Company

F. Budde, G. A. Farha, H. Frankemölle,
D. F. Hoffmeister, K. Krämer

**WILEY-VCH**

Weinheim – New York – Chichester – Brisbane – Singapore – Toronto

The Editors of this Volume

Florian Budde
McKinsey & Company, Inc.
Taunustor 2
60311 Frankfurt/M.
Germany

Gary A. Farha
McKinsey & Company, Inc.
600 14th Street, N.W.
Suite 300
Washington, D.C. 20005
USA

Heiner Frankemölle
McKinsey & Company, Inc.
Magnusstrasse 11
50672 Cologne
Germany

David F. Hoffmeister
McKinsey & Company, Inc.
600 Campus Drive
Florham Park, N.J.
07932-1046
USA

Konstantin Krämer
McKinsey & Company, Inc.
Taunustor 2
60311 Frankfurt/M.
Germany

■ This book was carefully produced. Nevertheless, authors, editors and publisher do not warrant the information contained therein to be free of errors. Readers are advised to keep in mind that statements, data, illustrations, procedural details or other items may inadvertently be inaccurate.

Library of Congress Card No.: applied for

A catalogue record for this book is available from the British Library.

Die Deutsche Bibliothek –
CIP-Cataloguing-in-Publication-Data.
A catalogue record for this publication is available from Die Deutsche Bibliothek.

© WILEY-VCH Verlag GmbH
D-69469 Weinheim, 2001

printed in the Federal Republic of Germany
printed on acid-free paper

Composition   K+V Fotosatz GmbH, D-64743 Beerfelden, Germany
Printing   Strauss Offsetdruck GmbH, Mörlenbach, Germany
Binding   J. Schäffer, Grünstadt, Germany

ISBN   3-527-30251-4

Preface
The Editors

These are troubled times for the chemical industry. After a hundred years as a major innovative force, the 1980s and 1990s saw it transformed into a mature industry with average profitability and low growth, no longer a favorite with investors.

Buffeted by the decline in demand stemming from the Asian crisis of the late nineties, the recent increase in the price of its major raw material – oil – and keen competition on the world's stock markets from New Economy companies, its management teams are now hard at work all over the world to regain attractiveness in the eyes of the shareholders.

However, they are finding that many strategy archetypes of the past are not producing tangible results. The overall attempt to move from cyclical to non-cyclical products and the general tendency to move downstream have had little effect. What is more, industry profitability has remained similarly unaffected by the implementation of quality management and the reengineering wave triggered by the implementation of ERP systems. New strategies are needed to return the sector to pre-crisis prosperity levels, and new routes must be sought to improve the productivity of the operational, marketing, and sales functions of chemical companies. Moreover, it is particularly crucial at this juncture to make these new approaches work. Unless management finds the right answers soon, the chemical industry runs the risk of being forever perceived as a typical commodity industry, with a flat development curve as the most realistic scenario.

The primary purpose of this book is to contribute to the discussion of the new paths, taking the issue of value creation as its guiding theme. As shareholder value is increasingly viewed as the key measure of corporate performance, we examine ways for chemical companies to attract and reward investor confidence. Our book summarizes insights that the global chemical practice of McKinsey & Company has gained in recent years in over a hundred engagements annually, working with virtually all of the leading chemical producers worldwide as well as private equity investors and other industry players. It aims to help managers identify areas where their companies could improve still further by describing the best practices currently applied in this changing environment.

But the book's intention also goes beyond this. Very few publications exist which comprehensively address the strategic management of chemical companies, and we are attempting to fill at least part of that void. It assumes a knowledge of business administration, without which it will not make easy reading. Nevertheless, we hope that it may be of interest not only to chemical industry managers, but also to many others who have dealings with the chemical industry, from students considering it as a career option to investors wishing to assess the value creation potential of a particular investment. Given that the chemical industry shares a number of common traits with industries such as pulp & paper, steel, or non-ferrous metals, it may also contain messages for managers in these industries, though not all of the examples from the chemical industry presented here will be transferable.

The book is organized in two parts. In the first section, we describe the relevance of shareholder value thinking and its implications for strategy creation in the chemical industry. Some particularly important recent trends such as e-commerce, biotechnology, and leveraged buyouts are analyzed in detail. To adapt freely from Thomas Alva Edison, however, only 10 percent of economic success is based on inspiration, and 90 percent on execution. The second part therefore deals at some length with the actions needed to make the strategies work and improve performance in general. The main thrust here is revitalization and change: how can chemical companies regain innovativeness? How can they put the right structures in place to achieve top performance? As traditional structures break down, how should chemical companies handle the moves towards more entrepreneurship in dealing with suppliers, and more individually tailored selling to customers? What consolidation options should they consider, and how should they ensure that the new entity created by a merger will operate effectively? Finally, we also consider the nature of the cyclicality that plagues the large commodity product sectors.

We do not claim to have covered all relevant topics. For instance, we do not discuss our very broad efforts to support the building of new businesses (apart from e-businesses). However, we decided that there needed to be a "design freeze" at some point, and we have done our best to cover most of the major issues for the industry in depth. The very important issue of environmental management is not included here since this is not our field, and can be better covered by others.

As far as we know, this is the very first publication of this particular kind which specifically addresses the chemical industry. We would therefore very much appreciate feedback from our readers. We have set up a dedicated e-mail address (*ChemBook@McKinsey.com*) for this purpose and look forward to your input, whether in the form of critique, suggestions or encouragement. It will certainly help us to gain an even better understanding of the current thinking and trends in the industry to which our professional lives are so closely bound.

About the Authors

McKinsey & Company, Inc., is an international top management consulting firm. Founded in 1926, McKinsey advises leading companies around the world on issues of strategy, organization, and operations, and in specialized areas such as corporate finance, information technology and the Internet, research and development, sales, marketing, manufacturing, and distribution. The authors are all members or alumni of McKinsey's client service staff. Most of them are affiliated to the global chemical practice, a worldwide network of consultants serving clients in this industry.

Markus Aschauer is an Associate Principal in the Paris Office and is a member of the chemical practice. He has served a large number of chemical clients on matters related to strategy, M&A, and operations. Markus has a PhD in industrial chemistry from the University of Milan. He has written a number of articles about the chemical industry with a focus on strategy and M&A.

Rolf Bachmann is a Principal in the Zürich Office and is a member of the leadership group of the chemical practice in Europe. Rolf has extensive experience in specialty chemicals and pharmaceuticals/life sciences. Besides wide-ranging Swiss and European business experience, Rolf has lived and worked in several Asian countries, including Singapore and Hong Kong. He is also a member of the European retail practice. Rolf has an MBA and a PhD in economics from the University of Zürich. Before his career with McKinsey, he worked for a major Swiss retail company in the area of logistics and support systems, and while completing his doctorate he also worked at the Institute for Swiss Bank Research.

Sönke Bästlein is a Principal in the Frankfurt Office and is a member of the leadership group of the chemical practice. His consulting work and knowledge development efforts are directed at micro-market management in specialties (including CRM and e-commerce), transformation of chemical companies towards service-based and asset-light businesses (including capital productivity), and M&A management. He leads the E&C (engineering & construction) sector and is the regional office manager of McKinsey in Frankfurt. Prior to joining McKinsey, Sönke worked for over five years with UHDE GmbH (now Krupp-Uhde-Engineering).

He has an MS degree in engineering from RWTH Aachen and a PhD in business administration from J.W. Goethe University in Frankfurt, where he was also a lecturer for three years.

Christophe Bédier is a Principal in the Paris Office, where he leads the French chemical practice and also co-leads the industry, electronics, telecommunications practice. He has worked extensively in the industrial and service sectors in France, Germany and the United States. Christophe is also a co-leader of the purchasing and supply management practice in Europe. He started his career with Thomson-CSF as an R&D project manager, then became a marketing product manager. He has also held the position of director of strategy in Rhône-Poulenc's chemicals division. Christophe holds an MS in engineering and an MBA from INSEAD.

Robert Berendes is a Principal in the Munich Office. He is a member of the leadership group of the chemical practice in Europe and of the German basic materials sector. He has worked on numerous studies in the chemicals, refinery, metals, and pulp & paper industries. His activities focus on strategy, organization and operational excellence. Robert holds a BSc and MSc in chemistry from the University of Cologne and a PhD in Biophysics from the Max Planck Institute for Biochemistry/Technical University in Munich. Prior to joining McKinsey, Robert worked at Procter & Gamble as a process engineer.

Florian Budde is a Principal in the Frankfurt Office and co-leader of the chemical practice in Europe. He has served a large number of chemical clients in strategic, organizational, and operational issues. He is also a member of the core group of the European corporate finance and strategy practice and has been involved in a number of corporate finance-related projects (divestment strategies, merger negotiations, target evaluation, etc.). Florian recently spent two years at McKinsey's Seoul Office. He holds a PhD in physical chemistry (surface science) from the Fritz-Haber-Institut der Max-Planck-Gesellschaft/FU Berlin in Berlin and has also worked as a researcher in the Physical Sciences Department of IBM Corp.'s T.J. Watson Research Center, New York. He has written a number of articles about the chemical industry.

Paul Butler is a Senior Expert in the London Office, and has served chemical industry clients around the world. His particular interests are process technology and economics, innovation, new materials and biotechnology. Before joining McKinsey, he worked for many years as a journalist covering the chemical, energy and engineering industries in the UK. He holds a BSc degree in chemical engineering from Birmingham University, and a PhD in solution thermodynamics from University College London.

Joël Claret is a Principal in the Geneva Office and is a member of the leadership group of the chemical practice in Europe. He has been involved in numerous organizational and strategy studies for major companies in Europe, the USA, and Japan, mainly in the specialty chemicals, energy, telecoms and electronic industries. He also has extensive experience with M&A, post merger, and change management topics. Before joining McKinsey, Joël held various positions at DuPont de Nemours Europe. He holds a degree in electrical engineering from the Swiss Federal Institute of Technology, Lausanne, and Carnegie Mellon University, Pittsburgh, as well as an MBA from INSEAD.

Jens Cuntze is an Engagement Manager in the Frankfurt Office and is a member of the chemical practice. His work has focused on managing costs as well as building businesses based on new technologies, mainly within the chemical industry. He holds a master's degree and a PhD from the Swiss Federal Institute of Technology (ETH) in Zürich as well as an MBA from the University of Wales/Fern Uni Hagen.

Christophe de Mahieu is a Principal in the Singapore Office and leads the chemical practice in Asia. He has wide experience in serving major chemical companies in both Asia and Europe in the areas of corporate and business development strategies, corporate finance, and operational excellence. Beyond chemicals, Christophe focuses on serving clients in the fields of energy and telecommunications. Prior to joining McKinsey, he worked with Exxon Chemical International in its Brussels and Houston Offices. Christophe has a business engineering degree and a master's degree from the Solvay Business School in Brussels.

Brian Elliott is an Engagement Manager in the New Jersey Office. His work has focused on the development of profitable growth strategies for advanced materials, chemicals, and industrial clients. Prior to joining McKinsey, Brian worked in research, development, and manufacturing for a variety of companies, including Allied Signal and Hoechst-Celanese, and for academic joint ventures in electronic packaging and utilities. He has a PhD in materials science and engineering from Northwestern University. His undergraduate degrees in mechanical engineering and materials science are both from Cornell University.

Philip Eykerman is an Engagement Manager in the Brussels Office. His recent work has focused on developing business unit/portfolio strategies for chemical companies. Prior to joining McKinsey, he was a lead process/project engineer with Fluor Daniel. Philip holds a chemical engineering degree from the University of Leuven and a master's degree in petroleum refining and gas technology from the Institut Français du Pétrole in Paris.

Khosro Ezaz-Nikpay is a Principal in the Cologne Office. He has done extensive work in purchasing and supply management, primarily in process industries (chemicals, pharmaceuticals, pulp & paper, and steel) and in the automotive sector. His other engagements have focused on strategy, post-merger management, pricing, and operational improvement issues. Prior to joining McKinsey, Khosro received a BSc from the University of California at Berkeley and a PhD in chemistry from Harvard University.

Gary A. Farha is a Director in the Washington D.C. Office and co-leader of the chemical practice in North America. Since joining McKinsey, he has worked on a variety of strategic, organizational, and operational efforts in North America, Europe, Asia, and Latin America. During the last few years, he has also led a number of practice building initiatives within the energy and chemical sectors. His research has been published in a number of industry periodicals. Prior to joining McKinsey, Gary held several financial positions with Kansas Gas and Electric Company. In addition, he worked for Boeing in supply management. He holds an MBA degree from Michigan State University and is also a chartered financial analyst.

Heiner Frankemölle is a Director in the Cologne Office. He is the global leader of McKinsey's chemical practice with about 200 consultants world-wide, and has worked in various fields of the process sector with a particular focus on specialty chemicals, petrochemicals, metals and energy. His main emphasis within McKinsey is on corporate transformation, strategy, M&A and post-merger management, as well as total operational performance and continuous improvement. He also recently led an internal effort to examine the strategic choices for the chemical players in the future and the impact on the industry landscape. He holds a PhD in agricultural economics (risk management) from the University of Bonn.

Boris Gorella is an Associate Principal in the Berlin Office and is a member of the chemical practice. He concentrates his efforts on e-commerce and post-merger integration projects in the chemical and processing industry. Prior to joining McKinsey, he worked as a Summer Associate at Goldman Sachs on M&A projects and as an Associate at A.T. Kearney on various strategy projects in the process industry. Before that, he worked at Rehau AG. Boris received a master's and a doctoral degree in chemistry from the Technical University of Berlin and an MBA from INSEAD.

Michael Graham is a Principal in the Brussels Office and is a member of the chemicals, energy, and business-to-business marketing practices. His work in the chemical industry includes the redesign of logistics and supply chains; value proposition re-engineering, and the redesign of core processes. Prior to joining McKinsey, Michael was an officer at Strategic Planning Associates (now a part of Mercer) in Washington D.C. He had previously worked for Polaroid in market research. He also taught for six years at MIT in materials science, macro-economic modeling and electrical engineering. Michael studied at the University of Leningrad and at Princeton. He holds an SB degree in mathematics, an SM degree in physics from MIT, and an SM degree from the Sloan School at MIT in international business and finance.

David F. Hoffmeister is a Director in the New Jersey Office and the leader of the chemical practice in North America. He has worked extensively with companies in pharmaceuticals, medical devices, chemicals and advanced materials in North

America, Europe, and South America, on areas including strategy, operations, and organization. In addition, David has been involved in a number of major post-merger management efforts in the chemical and healthcare industries. He has also recently led a number of internal McKinsey efforts to examine the evolution of the chemical industry and the impact on the various players. Prior to joining McKinsey, he worked for Chemed Corporation, a diversified specialty chemical and healthcare company and subsidiary of W.R. Grace, ultimately as the chief financial officer for one of the company's divisions. David received a BS in economics and finance from the University of Minnesota and an MBA from the University of Chicago.

Karsten Hofmann is an Associate Principal in the Frankfurt Office and is a member of the chemical practice. He is also a member of the European energy power & natural gas practice and the European leadership and organization practice. The focus of his consulting activities is on strategic, organizational and restructuring issues, predominantly in the chemicals and energy sectors. Karsten holds a PhD in organizational psychology from Mannheim University and worked for several years as a freelance consultant before joining McKinsey. He is the author of books on spans of control, upward feedback, and managing innovation.

Ulrich Horsmann is a Principal in the Cologne Office and is a member of the leadership group of the chemical practice in Europe. His consultancy work focuses mainly on strategy, organization, R&D/innovation management, and performance improvement issues in the chemical industry, both in Germany and internationally. Prior to joining McKinsey he was assistant professor of neurophysiology at the University of Cologne. He has worked in academia as a biologist, with the main emphasis on neurobiology and biochemistry. He holds a PhD degree in natural sciences/biology.

Helge Jordan was formerly an Associate in the Cologne office. He currently works as purchasing manager for the Vesuvius Group in England. He holds a master's degree in Chemistry from the University of Cologne and a PhD in industrial chemistry from the Swiss Federal Institute of Technology (ETH Zürich).

Konstantin Krämer is an Expert based in the Frankfurt Office and is a member of the chemical practice. Since joining McKinsey he has worked on a wide range of marketing & sales strategies and operations improvement issues in the chemical industry and related sectors. He is also in charge of the chemical practice infrastructure, including worldwide research and information services. Konstantin holds a master's degree in chemical engineering from the Technical University of Darmstadt. He received his PhD from the German Polymer Institute on the topic of plastics recycling.

Tomas Koch is a Principal in the Frankfurt Office and is a member of the leadership group of the chemical practice in Europe. His consultancy work includes a variety of strategic, operational, and organizational projects with a strong focus on post-merger management in Europe, the USA, and Asia. He has also led a McKinsey initiative on post-merger management in the chemical industry and has written a number of articles on this topic. Tomas studied economics and physics and earned a PhD in polymer physics from the University of Freiburg.

Simon Lowth is a Director in the London Office and is a member of the chemical practice. He has worked extensively on business strategy and operations improvement issues in the chemical industry and related sectors such as pulp & paper and glass. Prior to joining McKinsey, Simon worked as a mechanical engineer for Ove Arup & Partners, a major international engineering consulting firm, where he was responsible for the design and construction of process plant and offshore structures. Simon has an engineering degree from Cambridge University and an MBA from the London Business School.

Ralph Marquardt is an Associate in the Frankfurt Office and is a member of the chemical practice. His focus is on global strategy and e-commerce projects. Prior to joining McKinsey he gained three years of experience in the R & D department of Degussa AG, leading projects in Degussa's hydrocyanic acid derivatives and methylmethacrylate business units. He holds a PhD from the Ruhr-Universität Bochum.

Rob McNish is a Principal in the Washington D.C. Office and is a member of the leadership group of the North American corporate finance & strategy practice. He has served industrial clients in the areas of corporate strategy, alliances, and mergers and acquisitions. He is also a member of the chemical practice and has worked on issues of growth and business building in the chemical industry. Rob holds an MBA degree from the University of Chicago and is a chartered financial analyst.

David McVeigh is a Principal in the Stamford Office and is a member of the leadership group of the chemical practice in North America. He has served a large number of chemical clients on strategic, sales and marketing, operational, and e-commerce issues. In particular, he has extensive experience in polyurethanes, textile chemicals, food ingredients, adhesives, and various specialty and commodity plastics. He has also done extensive work investigating best practices in chemicals M & A and the future landscape of e-business in chemicals. David holds a BS in chemical engineering from Lafayette College, an MS in chemical engineering from Stanford University, and an MBA from Columbia University. Before joining McKinsey David worked at Air Products and Chemicals as a technical service and applications development engineer in polyurethane additives.

Jürgen Meffert is a Principal in the Düsseldorf Office and is co-leader of the global business building practice as well as co-leader of the German TIME sector (telecom, information technology, multimedia). He has worked on multiple issues such as marketing, pricing, strategy, operations, and M&A in the telecommunications, software services, computer and telecom equipment industries. He has also led multiple innovation and technology management efforts. Jürgen holds a degree in electrical engineering from Fachhochschule Münster, an MBA from J.L. Kellogg Graduate School of Management, Northwestern University, and a PhD from the University of St. Gallen, Switzerland. Before joining McKinsey, he worked for Nixdorf Computer AG in Germany and in the USA, and was on the Start-up team of the Open Software Foundation (OSF) in Cambridge, MA.

Rodgers Palmer is an Engagement Manager in the Washington D.C. Office. His major projects in the chemical industry have focused on identifying strategies for profitable growth, evaluating and redesigning a research portfolio, and providing options for improving a feedstock supply strategy. Immediately prior to joining McKinsey, Rodgers worked as a National Science Foundation Fellow in the Department of Chemistry at Yale University where his work focused primarily on the connection between hepatitis B and liver cancer. He has a PhD in molecular biophysics and biochemistry from Yale University and an AB in molecular biology from Princeton University.

Jan-Philipp Pfander is a Senior Engagement Manager in the Hamburg Office and is a member of the chemical practice. He has served the chemical and pulp & paper industries on a variety of strategic, operational, and organizational topics including major turnarounds, growth studies and mergers & acquisitions. Jan-Philipp is also a member of the European business to business marketing leadership group, where he currently focuses on e-commerce strategies and the application of e-commerce in marketing & sales across a range of industries. Jan-Philipp holds a master's degree in microbiology from the Max Planck Institute of Biochemistry/Technical University of Munich and a PhD in business administration from the University of Oldenburg.

Thomas Röthel is a Principal in the Cologne Office. He has also spent two years at McKinsey's Houston Office. He focuses his efforts very much on the chemical and utilities industries, being a member of both practices in Europe and leading the German energy sector. He works on strategic and operational issues and has extensive experience in the petrochemical industry. Thomas holds a PhD from the RWTH Aachen, where he worked on homogeneous catalysis.

G. Sam Samdani is an Expert in the New Jersey Office. He has worked with both chemical companies and leveraged buyout firms in assessing the attractiveness of opportunities in several major market segments. He also has a wide range of experience in strategic and operational issues. In addition, Sam has led and contributed to several internal knowledge-building initiatives, including the future landscape of e-business in chemicals and how to leverage the power of the Internet to add sustainable value to chemical businesses today. Previously an associate editor with *Chemical Engineering*, Sam has written extensively on best practices in the chemical process industries (CPI), and on emerging trends in CPI technologies and markets. Sam received his PhD from the University of Rochester and his bachelor's degree in chemical engineering from Yale University.

Wiebke Schlenzka is an Associate Principal in the Hamburg Office and is a member of the chemical practice and the pharmaceuticals/healthcare sector. She has worked on strategic, innovation related, and operational issues in the chemical, agrochemical and pharmaceutical sectors, with experience ranging across Europe, the USA, and Asia. She has also led a McKinsey initiative on biotechnology. Wiebke holds a PhD in molecular biology from the Biochemical Institute in Kiel and received a faculty award for outstanding research.

André M. Schmidt is an Engagement Manager in the Munich Office and is a member of the chemical practice and the metals & mining practice. Prior to joining McKinsey, he ran a freelance consultancy in the field of mobile computing in close relationship with Hewlett-Packard GmbH, Germany for more than ten years. André holds a master's degree in biochemistry from the University of Hannover and a PhD in protein chemistry from the Technical University of Darmstadt.

Christian Weber is an Engagement Manager in the Frankfurt Office and a member of the European corporate finance practice. He has worked in the transportation, financial services, electronics, electric utility, and retail industries, mainly focusing on corporate finance-related questions such as corporate strategy or transaction advice. He is also strongly involved in the development of the McKinsey knowledge base on capital market diagnostics. Christian holds a PhD in pure mathematics from McMaster University in Hamilton, Ontario, and worked at the Max Planck Institute for Mathematics in Bonn before joining McKinsey.

Acknowledgements

The editors would like to acknowledge all those who participated in the task of creating this material and preparing it for publication.

We are much indebted to the authors of the different chapters, who are core members of the leadership group of McKinsey's chemical sector, for their creative contributions and the effort they have made to put all of this together. We would also like to thank all other contributors to the work who were not directly involved in producing the written version.

A major vote of thanks also goes to the following members of our extended team: our manuscript editors Tim Hindle and Helen Robertson and our copy editor Ginni Light; Elisabeth Eliason who compiled the index; Andreas Appelhagen, Birgit Fehmel, Annett Jensen, Fleur Kennedy, Kimberly Kertis, Dee Anne Lamb, Jill Nathanson, Torsten Teichmann, Sari Varpa and Sandra Willis of McKinsey's Chemical Research and Information Service team; and administrative anchorwoman Anita Scheuermann.

Finally, may we acknowledge the biggest debt all: to those top managers and senior executives who contributed to our research in the various areas addressed here, and who also supported us with comments and criticism.

Florian Budde, Gary A. Farha, Heiner Frankemölle, David F. Hoffmeister, Konstantin Krämer

The Editors

Contents

1
Today's Chemical Industry: Which Way is Up?

Florian Budde and Konstantin Krämer

The chemical industry has emerged from a long struggle to sustain profitability, become leaner and more competitive, and use more sophisticated management approaches. Several eras in the industry's development have seen it change from exciting and innovative beginnings as the first ever science-based industry to its current maturity. Now it seems that a new era is dawning which will once more reshape the world of chemicals: the era of shareholder value.

1.1
The Chemical Industry Today – a Snapshot

Today, the chemical industry is one of the largest and most diversified in the world. The total value of chemical products sold in 1998 was about USD 1.5 trillion (including pharmaceutical sales of just over USD 300 billion). Western Europe accounted for 32 percent of this, the United States for 28 percent, and Japan for 13 percent (Fig. 1.1). Among the OECD member states, chemicals and petroleum products make a larger contribution to GDP than any other manufacturing industry. In Western Europe, chemicals account for approximately 2.5 percent of the total economy.

The chemical industry consists of hundreds of segments with an estimated 70 000 different product lines manufactured by more than 1 000 large and medium-sized companies, plus countless very small ones. The sector can be characterized as consisting of many "mini industries" of varying sizes, and it counts virtually every other industry among its customers – from agriculture to construction and electronics (Fig. 1.2). The automotive industry, for example, relies on several different chemicals in the production of tires, seats, dashboards and coatings, to name only a few major components.

Chemical products can be roughly segmented into basic chemicals, polymers, specialty and performance products, and agrochemicals. However, for practical purposes they can simply be classified as "commodities" and "specialties", the latter having higher value added and distinctive key success factors. Since chemical businesses are very heterogeneous, the range from "commodities" to "specialties" is a continuous spectrum.

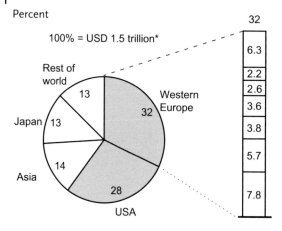

Percent

100% = USD 1.5 trillion*

Rest of world 13

Japan 13

Asia 14

USA 28

Western Europe 32

32
6.3
2.2
2.6
3.6
3.8
5.7
7.8

Fig. 1.1 Total chemical market volume by region, 1998
* Including pharmaceuticals, plastics and synthetic rubber, petrochemicals and derivatives, perfumes and cosmetics, paints and inks, inorganics, detergents and soaps, agrochemicals, dyes and pigments, fibers, fertilizers, other specialty chemicals;
Source: CEFIC – Facts and Figures, 1999

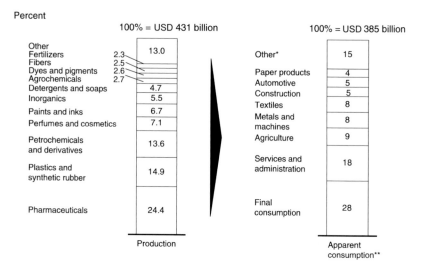

Percent

100% = USD 431 billion

	Production
Other	13.0
Fertilizers	2.3
Fibers	2.5
Dyes and pigments	2.6
Agrochemicals	2.7
Detergents and soaps	4.7
Inorganics	5.5
Paints and inks	6.7
Perfumes and cosmetics	7.1
Petrochemicals and derivatives	13.6
Plastics and synthetic rubber	14.9
Pharmaceuticals	24.4

100% = USD 385 billion

	Apparent consumption**
Other*	15
Paper products	4
Automotive	5
Construction	5
Textiles	8
Metals and machines	8
Agriculture	9
Services and administration	18
Final consumption	28

Fig. 1.2 Production and consumption of chemicals – Western Europe, 1998
* Includes pulp and paper, mining, electronics, textiles and refining; ** Apparent consumption (USD 385 billion)=production (USD 431 billion) – net trade balance (USD 46 billion); Source: CEFIC – Facts and Figures, 1999

In Western Europe and the United States, growth rates tend to be close to that of GDP – the minimum rate to maintain significance as an industry in the general economy. But there is significant potential for further growth in the emerging markets. In China and South East Asia, in particular, the industry's growth rates are considerably higher than the growth of GDP (Fig. 1.3). In Malaysia, South Korea, and Taiwan, for example, consumption of chemical products already accounts for over 12 percent of GDP.

The top players in the chemical industry are as global as any of their counterparts in other industries. Almost 40 percent of the sales of the top ten chemical

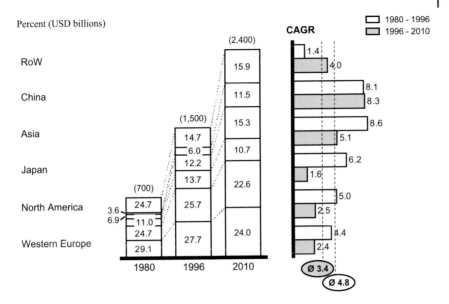

Fig. 1.3 Global chemical demand by region, 1980–2010
Source: Chemical Industries Association, Chemical Manufacturers Association, McKinsey analysis

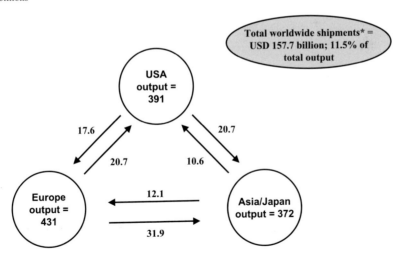

Fig. 1.4 Inter-regional chemicals trade, 1998
* Export values of the US (USD 51.8 billion), Latin America (USD 5.4 billion), Western Europe (73.4 billion), Asia/Japan (USD 22.7 billion), CEEC (USD 4.4 billion); Source: CEFIC – Facts and Figures, 1999

companies originate from overseas countries. The comparable figure for the top ten automotive and electronics firms is roughly 30 percent, and for the top pharmaceutical and oil companies around 50 percent. The industry has a distinctly regional character, however, in the sense that there are limited trade flows between the three main manufacturing regions (North America, Europe and Asia, including Japan). Only 11.5 percent of total output is shipped between these three regions (Fig. 1.4).

Not surprisingly, inter-regional trade is particularly limited for volume products, which are relatively expensive to transport. Nevertheless, the limited inter-regional trade is sufficient to couple prices and industry cycles in the different regions worldwide. The prices of basic commodity plastics, for example, moved in remarkably close harmony in all three regions for the last two decades of the twentieth century.

In most places, however, trade within the regions is very strong. For instance, 51 percent of Europe's total chemical output is exported within the region (including Central and Eastern Europe); 17 percent of it is exported outside Europe; and only 32 percent is consumed domestically.

Germany and the United States are the world's largest exporters of chemical products. In 1998, Germany exported USD 69.4 billion's worth and the United States USD 68 billion's worth (Fig. 1.5). These two countries are also by far the biggest net exporters of chemical products.

Trade patterns for chemical products differ markedly from those of overall trade. Countries like Taiwan, China, and South Korea, for example, which have large surpluses in their overall trade account, have sizeable deficits in their trade in chemicals. The United States and Britain, on the other hand, have large trade deficits overall, but sizeable trade surpluses in chemicals.

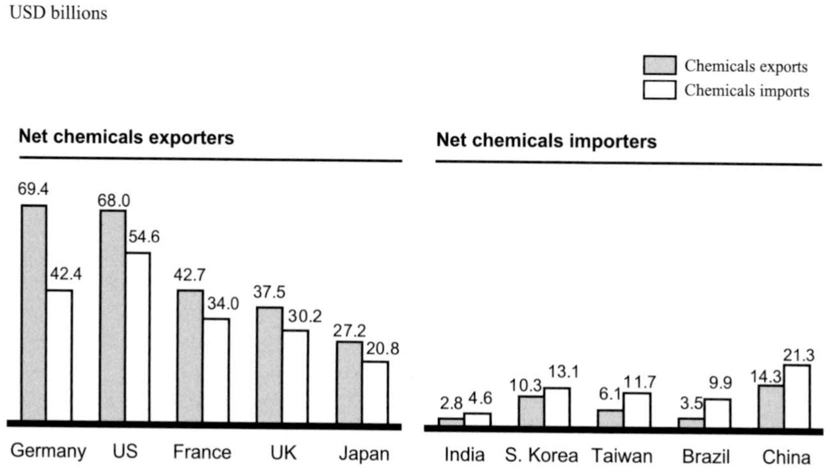

Fig. 1.5 Chemical import/export balance of major economies, 1998
Source: Chemical Manufacturers Association

Percent

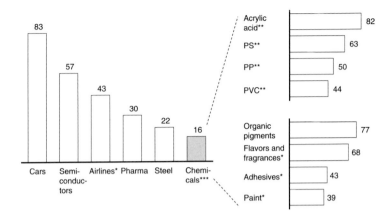

Fig. 1.6 Concentration level of the chemical industry – top ten worldwide market share, 1998
* 1997 data; ** 1999 data; *** Without pharmaceuticals; Source: McKinsey analysis

One interesting aspect of the chemical sector is that the industry as a whole is highly fragmented. The top ten companies in chemicals (excluding pharmaceuticals) account for only 16 percent of the total market, well below other industries, such as automobiles, where the top ten firms account for 83 percent of sales, or semiconductors, where the top firms account for more than half of all sales (Fig. 1.6). On the other hand, at the product segment level, the top ten manufacturers of acrylic acid, for instance, account for 82 percent of their market. The top ten manufacturers of organic pigments account for 77 percent of their market, and the top ten in flavors and fragrances for 68 percent.

The level of concentration in the industry also varies by region, with North America in general showing the highest concentration and Asia the lowest. For acrylic acid, for example, the top four manufacturers in the United States account for the whole market. In Western Europe they account for 98 percent but in Asia (excluding Japan) they account for only 76 percent of the market. The comparable figures for the top four producers of PVC in the United States, Western Europe and Asia (excluding Japan) are 78 percent, 58 percent and 45 percent respectively.

There are plenty of other examples (e.g., polypropylene and polystyrene) where the Herfindahl Index, a measure of industry concentration frequently used by antitrust authorities, is rather high in North America and Western Europe. This necessarily limits the opportunities for Western players to grow by means of mergers and acquisitions in their own domestic markets. Given this structure, it is not surprising that Western players are becoming increasingly interested in acquiring Asian companies. In addition, the large number of Asian players in several industry segments often results in sub-scale plants with fairly inefficient operations. The average plant size of an acrylic acid manufacturer in Asia (excluding Japan), for instance, is less than one-seventh of the average size of a corresponding plant

Changes in line-up of chemical players			Major M&A changes in 1998/99
1980	**1998**	**1999**	• Hoechst and Rhône-Poulenc spun off their chemicals businesses (Celanese and Rhodia) and merged the life science parts into a new entity (Aventis)
1. Hoechst	• BASF	• BASF	
2. Bayer	• DuPont	• DuPont	
3. BASF	• Bayer	• Dow/Union Carbide*	
4. DuPont	• Dow	• TotalFinaElf	
5. ICI	• Shell	• Bayer	• Mergers of oil companies
6. Dow	• ICI	• ExxonMobil	– Exxon and Mobil
7. Union Carbide	• Hoechst	• ICI	– Total, Petrofina, and Elf
8. Shell	• Exxon	• Shell	– BP and Amoco
9. Exxon	• Rhône-Poulenc	• Akzo Nobel	• Merger of chemical companies
10. Montedison	• Elf Aquitaine	• BP Amoco	– Dow and Union Carbide

Fig. 1.7 Changes in top ten chemical companies
* Pro forma for merged entities; Source: McKinsey analysis

in North America. Therefore, the pressure on Asian players to further consolidate their operations will continue.

The top ten firms in the industry are all Western, and for the best part of the last two decades of the twentieth century there was little change in the names appearing in annual lists of the top ten. Eight of those in the 1980 listing, for example, also appeared in 1998 (Fig. 1.7).

However, between 1998 and 1999, some dramatic shifts occurred as the result of a number of large mergers and acquisitions. On the one hand, Hoechst and Rhône-Poulenc spun off their traditional chemicals businesses (to become Celanese and Rhodia) and merged their life science divisions into a new entity (called Aventis). On the other hand, several other companies – mainly oil-based – merged their businesses in order to gain economies of scale. The result was that several new names appeared among 1999's top ten: BP Amoco, Dow/Union Carbide, ExxonMobil, and TotalFinaElf.

Despite the regional structure discussed earlier, all the top ten giants are truly global companies with large chunks of their business in each of the three main regions. In addition, the portfolios of products manufactured by these giants are very diverse, with the major conglomerates ranging from primarily commodity players to hybrids which operate at all levels of the industry, with only one of the group focusing entirely on specialty chemicals (for details, see Chapter 3).

There are management concerns in each of these major categories. Commodities have no real competitive differentiation, and many specialties segments are too small and fragmented, with growth rates which have come down substantially over the last five years. The management of hybrids might find it difficult to be successful across a very wide range of products and markets. In Chapter 3 we will return to these issues and describe the potential strategic choices for the chemical industry in the new millennium.

1.2
Eras of the Chemical Industry

Early chemical companies can be compared to the software giants of the nineties: new, innovative, and completely revolutionary. The industry remained exciting and attractive until as recently as the last ten to twenty years, but with increasing maturity and declining innovative potential it has come to be regarded as somewhat dull, a less attractive investment or employment prospect. What, then, does the future now hold?

It may be useful to set the scene for our analysis of the chemical industry's current situation and future challenges by taking a brief look at its evolution and, in particular, by outlining our understanding of the changes in general industry dynamics in the various chemical businesses over the past decades and up to the present day.

Industrial historians trace the origins of today's chemical industry back to the discovery in 1856 of a synthetic process for manufacturing mauve dye. Before that date, chemical production was largely confined to the unsophisticated processing of inorganic compounds found in the ground, and was little more than an offshoot of mining. The discovery of this organic dye, by the Englishman William Henry Perkin, marked the beginning of the world's first science-based industry.

From the moment of Perkin's discovery until today, the development of the chemical industry falls into six major eras, during which it has been transformed

	Foundation	Development	Expansion	Diversification	Maturity	Shareholder value
Scope	• Scientific discoveries • New molecules	• Upscale from laboratory to production • Emergence of polymers	• Substitution • Internationalization	• Broadening of product range	• Global competition • Restructuring	• Focus on financial returns • Further disaggregation
Key factors of success	• R&D • Access to raw materials	• Process development • Engineering • R&D	• Process technology • Integration • International presence • R&D	• Sales • Marketing • Process and application technology	• Operational excellence • Economies of scale • Market conduct	• Investor relations • M&A capabilities • New business generation
Growth relative to GDP	• Low	• Medium/high	• High	• Medium/high	• Medium	• Medium/?
Industry structure	• National, fragmented	• National, multidivision companies	• Appearance of multinationals	• Move to conglomerates	• Consolidated multibusiness chemical conglomerates	• ?
Timeline	1920	1950	1970	1985	1995	→

Fig. 1.8 Era analysis chemical industry
Source: McKinsey

into the world's largest manufacturing industry, an industry now indispensable to almost every manufacturing process (Fig. 1.8).

Foundation: In this era the chemical industry was putting down its roots. As the industrial revolution created demand for many chemical products such as dyestuffs, this period was characterized by the discovery of new molecules, based for the most part on different types of hydrocarbons. In addition to R&D, the access to raw materials was the key success factor. With the development of the coal industry in the nineteenth century discoveries were plentiful. This was the starting point of many national chemical companies, which still exist today as global players. Production during this period, however, was small-scale and fragmented.

Development: The industry then moved out of the laboratory and into the factory. The emphasis shifted from discovery to large scale organized production. Engineering and process development became as important as chemical R&D. The Haber-Bosch ammonia process was the leading example of a catalytic process operated on a large scale, requiring important advances in process design and engineering. Companies in the industry began to develop more sophisticated multi-divisional structures, and growth began to accelerate. One important reason for that was the emergence of polymers: most of the modern mass polymers like polystyrene, PVC, polyethylene, and polyurethanes were discovered between 1930 and 1940.

Expansion: Chemical companies grew rapidly as a result of the increasing substitution of synthetic products for natural materials. Plastics, in particular, replaced natural products, such as paper, wood, or cotton, in many applications as they were cheaper and easier to process. Based on the importance of crude oil as a raw material for nearly all organic chemicals, large integrated chemical complexes like those in the North American Gulf Coast region or Rotterdam and Antwerp in Europe were established, and economies of both scale and scope became important. Another major trend was that the chemical companies began to spread outside their home markets. At the same time, true multinational organizations began to emerge, as international presence became as important to success as scientific discovery and process development.

Diversification: Companies broadened their product ranges even more widely. The key to success here switched from R&D to processes and application technology as well as marketing and sales. This change resulted in the build-up of strong customer service and technical support departments. This was the stage when many players transformed into chemical conglomerates with highly diversified product portfolios ranging from commodities through specialty chemicals to pharmaceuticals. The industry's growth rates – still well above GDP – decreased, however, after the oil crises (1973 and 1979).

Maturity: The creation of new products slowed down and the industry became more focused on global competition between firms. It was harder for companies to grow faster than GDP, and they were forced to achieve operational excellence through aggressive restructuring and cost control. The number of mergers and acquisitions began to accelerate in order to consolidate the industry and to achieve further cost synergies through economies of scale.

Shareholder value: In this most recent phase the main focus is on financial returns, that is on the value created for investors. M&A activity in the industry is widespread but disaggregation is also a feature of the corporate landscape as more and more firms choose to focus on a smaller number of fields. The need for value creation imposed by capital markets is forcing Western chemical players to take action, which is sometimes quite painful. They have had to fundamentally rethink their strategies and ways of doing business. In particular, the process of building new businesses outside the traditional scope of activities has provided an opportunity to crack the growth challenge (see also Chapter 3).

Obviously, these six eras have occurred at different times in different parts of the world. In addition, the degree of participation has been different. In the early years, it was Europe that led the way. Britain and Germany, with the USA joining in from the 1920s onwards, mainly drove the foundation and development phase of the industry. The expansion, diversification, and maturity eras took place in Europe and the United States at more or less the same time. They occurred in Asia rather later and somewhat less intensely. The era of shareholder value, however, came first and fastest to the Anglo-Saxon world. European firms have only just begun to follow their US counterparts into this phase. In Asia, the industry has not yet really begun to move into the shareholder value era.

To understand this recent evolution and the perspectives for the chemical industry it is essential to understand the capital market point of view, which is driving the strategies of most Western chemical players today. The importance of the shareholder value era for the chemical industry is discussed in the following chapter.

2
Today's Challenge – Value Creation
Robert Berendes, Florian Budde, Konstantin Krämer, and Christian Weber

Value creation is now the key ratio for performance, and has to become the main focus for all companies seeking financing in the ever more fiercely competitive international capital markets. Given the slowdown in growth for a large part of the industry, tough competition, and growing numbers of market players, top management in the capital-intensive chemical industry is facing increasingly tough challenges in creating value. We will describe here how pursuing shareholders' best interests necessarily and logically involves pursuing the best interests of all stakeholders, a point of view which is also supported by empirical evidence. We will also explain a powerful set of tools which is available to help top management in formalizing its approach to managing capital market expectations, and point out some specific areas in which companies have to perform well to gain shareholders' confidence.

2.1
Shareholder Value – Why the Recent Rise to Fame?

Putting the owner first: it seems like a blindingly simple idea. Yet the notion that the fundamental goal for management should be to maximize the organization's value to its shareholders has only really caught on in recent years, ousting the shorter term focus on Return on Sales (ROS). Already widely accepted in the Anglo-Saxon world, it is now rapidly gaining ground in continental Europe, and is starting to become a priority in Asia.

Why is this concept only currently winning support after so many years? There are three main reasons:

The *first* reason is that shareholder value is growing in strategic significance. With the globalization of capital markets has come the free flow of equity across borders. Added to this, the unending search for economies of scale has led to increased consolidation in industries across the board – from capital-intensive ones like chemicals and mining to more knowledge-based businesses like pharmaceuticals and financial services. As a result, any company failing to create shareholder value is likely, sooner or later, to become a potential target because of its relatively low market capitalization.

The upside of this is that companies whose share-price performance continually outshines their rivals' are able to use their own equity to fund their acquisitions, increasing their market power by getting competitors under their control.

The *second* reason for the central significance of shareholder value is the growing realization that, in the long run, what is best for shareholders is best for everybody else too. In other words, there is a congruence between the creation of shareholder value and the interests of the other stakeholders in an organization.

Many major investors, such as pension funds, some mutual funds, and – in some cultures – banks are in favor of long term security. They are looking to their investments to produce long term growth in value and are therefore unlikely to compromise the interests of other stakeholders who contribute to making companies successful. Furthermore, a poor shareholder value performance makes companies unattractive to top talent and other stakeholders, undermining their competitive position in the long term.

Of course, there is always a possibility that productivity increases will be achieved in the short term at the expense of labor. In the medium term, however, the most productive companies will gain market share and will, as a consequence, increase their level of personnel to above average rates. In other words, winning companies seem to create greater value in relative terms for all stakeholders – be they customers, employees, the government, or providers of capital. When winning companies are compared with their competitors, they show higher productivity and greater increases in shareholder wealth as well as higher employment (Fig. 2.1).

Thus, beyond the short term, all those who have a claim on the organization benefit when shareholders (or their agents) set out to maximize the value of their own claim. In the attempt to maximize value for themselves, shareholders cannot help but maximize value for others.

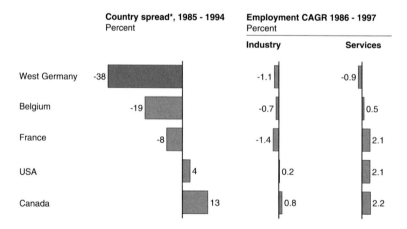

Fig. 2.1 Higher productivity correlates with higher employment level (long term)
* (ROIC-WACC)/WACC; Source: OECD (Labour Force Statistics 1998 edition, March 1999), McKinsey analysis

USD thousands per 1%

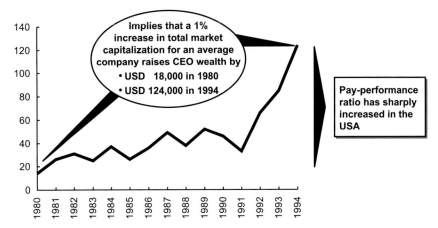

Fig. 2.2 Total change in CEO wealth relative to annual shareholder return
Source: B. Hall, J. Liebman, "Are CEOs really paid like bureaucrats?", Quarterly Journal of Economics (1998) 3:653–691

The *third* reason why management is increasingly focusing on shareholder wealth creation is that its own income has become more linked to share price development, for example, through the use of stock options (Fig. 2.2).

In addition, increasing investor power – due to a growing share of private shareholders and bundling of their interests in mutual and pension funds – amplifies the above reasons: more powerful investors will be less willing to cope with a below average return and will as a result be more willing to abandon an underperforming stock. This will intensify the pressure on the stock price.

Although the creation of shareholder value is becoming the primary target for top management in all kinds of industries, the level of acceptance varies widely. In some more traditional industries, like steel and chemicals, senior managers still tend to run their business with a focus on the Return on Sales or even tons of output. Yet this ratio has a major flaw: it fails to judge the company's results in relation to its invested capital and the cost of that capital, which could be fatal in the longer term. However useful ROS may be, it cannot stand alone as a measure of performance.

Since the mid-nineties, more and more chemical companies have come to realize the importance of shareholder value, and it is crucial for this development to continue. In such a capital-intensive industry, the lack of focus on value creation (and the use of ROS as the key ratio) will typically lead to relatively low returns on invested capital. Ultimately this creates a situation in which even "successful" players do not generate value and, in some cases, actually destroy it; the typical spread (i.e., the difference between return on invested capital and capital costs) for chemical companies lies between –3 percent and +5 percent, with only a few exceptions.

2.2
How Good a Measure is Shareholder Value?

Given the occasional wild swings on the stock market, does the market price of an investment really reflect an unbiased estimate of the true value of that investment? In addressing this question, we should perhaps start by defining what we mean by the terms.

The stock *price* of a company is the result of short term clearing of a supply-demand situation on the stock market, and is influenced by day to day fluctuations: as in other market situations, the price and the quantity sold are determined by the slope of the supply and demand curves.

The fundamental *value* of a company, on the other hand, is the net present value of the expected future cash flows, discounted by the cost of capital (DCF). This only alters for the better or the worse if fundamental changes occur, for example if prices change, new technologies are introduced or the company achieves a breakthrough into new markets.

Logically speaking, the two should match up; and, over the long term, we find that they actually do. It cannot be denied that it does not always look that way. It is the wild fluctuations, booms and busts and sudden and unexpected collapses in individual share prices that catch the public's attention. Granted, there are cases where profit warnings take markets by surprise. There are also times when they appear to neglect announcements of good results or strategic moves such as acquisitions.

Capital markets may not always be efficient on an "absolute scale". However, at any given moment they are the most efficient reflection of what the most likely future development could be, given the information available at that time. In this context, a number of typical prejudices about capital markets have been proven untrue by academic research.

The market takes the short term view. No. Capital markets are normally able to predict future performance very well. Typically, share-price movements anticipate changes in a company's earnings about one to three years in advance. Figure 2.3 shows the lag between the fall in IBM's share price (relative to the S&P 500 index) and its EPS in the late 1980s and early 1990s. The market was signaling the company's decline well in advance of its appearance on an earnings statement. Empirical evidence also shows that there is a high correlation between the DCF and the market values of companies from different industries. Stock market valuations over time are amazingly close to the DCF – when applied to large companies which are considered to be completely transparent and have a long-standing track record (a typical enough profile in the chemical industry).

Stock market fluctuations cause major distortions in the relative valuations of companies. Not so. Even in highly turbulent markets, the relative performance of stocks remains remarkably stable. For example, both before and after the 1987 stock market crash, the ratio of the market values of pairs of companies from the same industry (such as General Electric and Siemens, DuPont and Bayer, and General Motors and Volkswagen) did not change by more than 10 percent, despite the steep fall in the total market valuation.

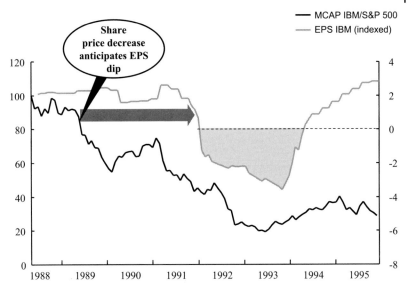

Fig. 2.3 Market capitalization compared to S & P 500 and EPS of IBM
Source: Datastream, McKinsey analysis

Markets can be influenced by earnings increased on the basis of changed accounting policies. Again, this is not so. Markets are not fooled by the mere tinkering of accountants. Consider a couple of typical examples: in one case, a company switched from an accelerated to a straight-line method of depreciation for reporting purposes only. The switch resulted in higher reported earnings but left cash flows unchanged, and it was not accompanied by any abnormal improvement in the company's stock-market performance. In the second case, a company's switch from a FIFO (first-in-first-out) to a LIFO (last-in-first-out) inventory policy brought down its declared earnings but increased its cash flow. The switch led to a significant positive reaction from the stock market.

The market is slow to react. No, between 75 percent and 90 percent of the stock-price adjustments that take place after a surprising announcement on earnings occur within the first three hours of the relevant announcement. This is true of both favorable and unfavorable messages, though there are some exceptions here.

The market's predictions on whether mergers and acquisitions will be successful are based on wild guesswork. In fact, the reverse is true. Empirical evidence shows that the expected value creation of a merger (as measured by the Total Return to Shareholders (TRS) of the relevant companies in the period from five days before the announcement of the deal to five days after) correlates strongly with the actual value realized, as measured by the TRS of the relevant companies over a period of several years, discounting the value creation of the market as a whole over the same period.

Buying into the paradigm that in the long term value equals price, in other words, there is no such thing as an under- or overvalued company, has interesting consequences for the management of a company: stock prices then become a reflection of current capital market expectations of a company's future economic performance. The management therefore needs to understand and actively manage these expectations.

Several tools have been developed to clarify and manage the link between a company's strategy and its performance on the capital markets. (For a detailed overview please see Copeland/Koller/Murrin, "Valuation", 3rd edition, John Wiley & Sons, Inc., 2000.)

2.3
Understanding and Managing Capital Market Expectations

There are essentially three areas where the link between a company and the capital markets can be analyzed and quantified: understanding historic capital market performance, quantifying and decoding current market expectations, and assessing the value impact of strategic actions.

2.3.1
Understanding Historic Capital Market Performance

While – as seen above – TRS is a good measure for any company to optimize, it would be unfair to blame (or give credit to) management for the whole of a company's TRS performance. No company can free itself entirely of the general influences on industries, national economies and the world economy which will affect its performance on the stock market.

However, breaking down TRS into its component parts not only helps to explain share movements, but also allows companies to analyze their own underlying performance and benchmark themselves against competitors. We call this the DEEP (Delta in Expectations of corporate Economic Performance) model, after the key component of TRS for management purposes – and the one that is most open to management influence (Fig. 2.4).

The Total Return to Shareholders for a particular period can be broken down into three main elements:

- The first is the Cost of Equity. It is defined here as the return that a shareholder expects from the company over a certain period, in terms of dividend and the capital gain from a rise in the stock price. These are the actual expectations of income on which an investor bases his original purchase or reviews his portfolio. The Cost of Equity can be estimated using the Capital Asset Pricing Model (CAPM).
- The second element consists of unexpected changes in financial markets generally, reflecting broader economic shifts such as changes in interest rates or in the expected risk premium during the period under consideration. These effects could not be predicted by the stock buyer at the beginning of the period, and are obviously beyond management's control. They also clearly influence all stocks of a similar type.

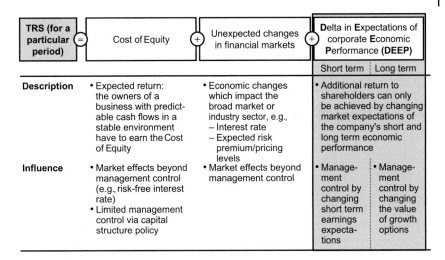

TRS (for a particular period) =	Cost of Equity +	Unexpected changes in financial markets +	Delta in Expectations of corporate Economic Performance (DEEP)	
			Short term	Long term
Description	• Expected return: the owners of a business with predict-able cash flows in a stable environment have to earn the Cost of Equity	• Economic changes which impact the broad market or industry sector, e.g., – Interest rate – Expected risk premium/pricing levels	• Additional return to shareholders can only be achieved by changing market expectations of the company's short and long term economic performance	
Influence	• Market effects beyond management control (e.g., risk-free interest rate) • Limited management control via capital structure policy	• Market effects beyond management control	• Manage-ment control by changing short term earnings expecta-tions	• Manage-ment control by changing the value of growth options

Fig. 2.4 Framework for TRS decomposition
Source: McKinsey

- The third element is what we call the Delta in Expectations of corporate Economic Performance, or DEEP. In simple terms, this is the value impact of the difference between the expectations at the beginning and at the end of the period considered, after adjusting for the changes in financial markets. In other words, it is the component which reflects the difference between good performance and bad, because it is the part which reflects whether or not management has done better or worse than the informed investor could have expected, or indeed has performed in line with expectations. Moreover, since it is the measure of management impact, this element can be considered to be fully under management's control. DEEP can be broken down into two parts:
  - Short term: the part due to changes in short term (1–3 years) earnings expectations
  - Long term: the part due to changes in the company's price level (P/E ratio). Typically, changes here reflect the market perception of the company's ability to grow its business organically.

With value creation as the key measure of corporate performance, it is clear that TRS beyond the Cost of Equity and overall changes in financial markets can only be achieved by changing the market's expectations about a company's short and long term economic performance, because, as we have seen, these are the only fully influenceable parts of the share price. Both elements of DEEP are under the full control of the company's management: the short term one by changing the profitability of existing assets; the long term one by changing the value of growth options.

As a management tool, DEEP offers two major benefits. First, by tracking its own DEEP over time, a company can compare how strategic actions taken at the beginning of a period influenced the DEEP in this period, a test of the market's perception of the company's strategy. Second, by comparing its DEEP with that of other companies in the industry it can gain a clear idea of its relative performance – and again, failure to match up to best performance can provide an incentive to seek improvement opportunities. The particular benefit of the DEEP over other ratios here is that it strips out the part under management control.

2.3.2
Quantifying Future Expectations

Under the assumption that today's stock price is equal to the present value of future cash flows, managers can perform a set of analyses to understand the market's growth and profitability expectations by using a simple valuation model. By comparing the expectations implied by the share price with their own business plans, management might uncover a need for action. From the many shades of gray that could result from such an analysis, let us outline two extreme cases:

"Edge of the cliff": the expectations of growth and profitability implied by the company's share price are far beyond the best case scenario in the company's business plan. That means – assuming that the market will find out at some point – that the company is facing a revision of expectations, that is negative DEEP or underperformance in TRS. It could consider remedial action in its activities to prevent this; but also, in the short term, it could use its inflated stock as acquisition currency.

"Mountain of gold": the market's expectations of the company are very low, even much lower than the worst case scenario in the company's business plan. That means that once the company has managed to convince the market of its "real" future economic performance prospects, it will have a positive DEEP, that is, TRS outperformance. Management teams in this situation would be well advised first to reconfirm their planning to make sure that it is truly realistic, and second to put an effective investor communication strategy in place.

One very simple analysis for assessing future expectations is to calculate the split of a company's share price into the part which will be derived from the business as it is at present and the part which will come from organic growth. The value of the "business as is" can be interpreted as the (hypothetical) value of the business if all additional investments were stopped and all earnings were distributed to the owners. The – usually positive – difference between this value and the company's share price can be interpreted as the value of the company's growth options (Fig. 2.5).

The next level of sophistication can be reached by reverse engineering the company's share price in terms of growth and capital efficiency (e.g., ROE). It might be noteworthy that a higher growth only results in higher value if the return on capital exceeds the cost of capital, that is, ROE higher than the Cost of Equity or ROIC exceeding the Weighted Average Cost of Capital for both equity and debt (WACC). Mapping market expectations about the growth and profitability of a company and of its competitors can provide insights about potential "edge of the cliff" or "mountain of gold" areas.

Percent

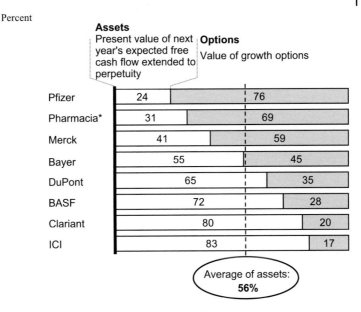

Assets
Present value of next
year's expected free
cash flow extended to
perpetuity

Options
Value of growth options

Pfizer	24	76
Pharmacia*	31	69
Merck	41	59
Bayer	55	45
DuPont	65	35
BASF	72	28
Clariant	80	20
ICI	83	17

Average of assets:
56%

Fig. 2.5 Short and long term portion of share price for selected companies
* Including acquired pharmaceutical business of Monsanto; Source: Datastream, McKinsey analysis

2.3.3
The Impact of Strategic Action

Chemical companies will attain high market values only if they are able to play on market expectations to the fullest extent. Success here depends on superior execution of three major interdependent elements:

1) Companies have to ensure that they build management credibility in the eyes of the capital markets by avoiding under-delivering against short term earnings expectations and, if at all possible, over-delivering from time to time. This will result in a higher share price at a constant P/E ratio.
2) On top of that, they have to create new viable growth options. Credible moves in this area will potentially lift market expectations to new levels, resulting in a higher P/E ratio.
3) Last, and not to be underestimated, proactively influencing investor expectations through transparent and reliable communication backed by facts and actions can have a major impact.

2.3.3.1 Delivering on Short Term Earnings Expectations
As we have discussed, stock markets take the long view. Despite this, however, it is also critical to build the confidence of the investor community by avoiding under-delivering (indeed, over-delivering wherever possible) in the short term against

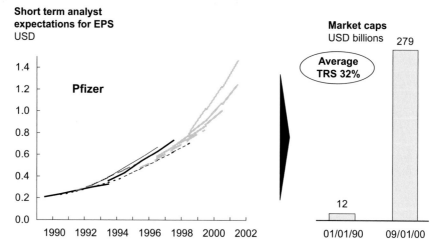

Fig. 2.6 High performers achieve credibility by raising EPS and by fulfilling or over-delivering against expectations
Source: Datastream

the market's expectations. Best practice here is a continuous growth in earnings per share (EPS) over time, uninterrupted by any significant intermediate profit dips. The US pharmaceutical company Pfizer is a great example of how to gain the confidence of investors by continuously increasing EPS over many years, with measures including a major restructuring project in the early nineties (Fig. 2.6).

By contrast, the initial credibility that is built up by short term boosts in EPS can rapidly disappear if the growth in profitability starts to slow. A leading sports-wear producer, for example, lost the confidence of investors when its profit growth started to slow down in the late 1990s following an initial boost in its EPS some years earlier.

But even after an intermediate dip in its EPS, a company can regain investors' confidence by turning its performance around rapidly and rejoining a path of continuous growth in profitability. The US chemical giant DuPont is a good example of that. The decline in its EPS led to a massive restructuring effort in the mid-nineties which formed the basis for a healthy growth in profit in the late 1990s.

Of course, the benchmark will vary for different sectors of the chemical industry. For example, it might be fine for a highly capital-intensive producer of basic chemicals to earn just above its cost of capital. But this benchmark would clearly not be sufficient for a top player in the industry with significant business in more knowledge-driven segments, such as pharmaceuticals or specialty chemicals. Here, a company with high aspirations should benchmark itself against pure pharmaceutical players achieving ROIC (Return on Invested Capital) values that are at least 20 percentage points higher than their cost of capital.

It is also crucial for chemical companies to constantly lift (and then fulfill) the expectations raised by these benchmarks. Failing to meet EPS targets on a regular basis is punished by the markets. The market capitalization of companies guilty of this is lower by a factor of two or more than that of other industry players.

2.3.3.2 Creating a Long Term Growth Story

Superior market capitalization is achieved faster through the creation of new viable growth options that lift the market's previous expectations to new levels, potentially leading to a significant increase in the company's P/E ratio. It appears that the really successful companies have to be prepared to step out into new strategies and new arenas in building credible growth scenarios (see Chapter 3). Winning plays can also include the announcement of a merger, the launch of a blockbuster or high-performance product, or a significant change in the top management – for example, the announcement of the appointment of a new CEO with a successful track record. However, it is not the one-off event that drives the share price increase, but the market's belief in the additional growth potential implied in this discontinuity that makes the big difference.

Interestingly, in a comparison of excellent companies with those that perform less well (in terms of their capital market performance), the excellent companies achieved superior returns for shareholders not by avoiding bad years, but rather by achieving really good years – that is, years in which they statistically outperformed the industry index by a significant margin – more than three times as often as the average companies.

2.3.3.3 Effective Communication

Good communication with the investment community is essential for a company that wishes to increase its market value. Our analysis of a sample of German companies found a significant correlation between the quality of investor relations (judged by the quality of annual reports, analyst conferences and other investor events) and shareholder returns. The sample covered a broad spread of industries and included several chemical companies.

Investors are particularly annoyed by negative surprises – about, for example, quarterly earnings. The share price of a major car manufacturer dropped by about 9 percent compared with the market overall in the three days following a negative earnings announcement by the company. Analysts at the time said that about one-third of the drop was purely due to bad communication.

In a nutshell, then, if a company's top management wishes to obtain superior returns for its shareholders, it has to do three things in particular:

1) It has to generate a story of continuous earnings flow through strategic and organizational direction-setting;
2) Furthermore, it has to create a compelling growth story about the future;
3) Finally, it has to communicate with the investment community proactively and reliably.

3
Strategic Choices for the Chemical Industry in the New Millennium

Florian Budde, Brian Elliott, Gary A. Farha, Heiner Frankemölle, Tomas Koch, and Rodgers Palmer

In defiance of its comparatively advanced position in the life cycle of industries, the chemical industry managed to sustain market average performance in terms of profitability, growth, and value creation right up to the late 1990s, outdoing many comparable mature industries. However, ever since the Asian crisis, the chemical industry has been lagging behind the general markets. To return to pre-crisis levels, chemical companies will need to improve performance significantly. Increasing focus through restructuring and consolidation is one successful route, although it has limitations. At the same time, companies should be leveraging the most important part of their capital – their knowledge. By combining a new set of non-traditional knowledge-leveraging strategies with the more traditional ones, companies can secure profitable growth and above average value creation.

3.1
The Chemical Industry Today – Forces for Change

In general, the industry's players fall into three main categories as described in Fig. 3.1. The first group, accounting for about 36 percent of total sales, focuses on commodity chemicals (single molecules sold on the specificity of the molecule, differentiated only by their purity level or impurity profile) and includes companies such as Shell Chemicals and Union Carbide. The second group focuses on specialties (single molecules or formulations which are sold on the basis of functionality or performance) and includes companies such as Ciba Specialty Chemicals and Rohm and Haas. The final group consists of the hybrids, large chemical conglomerates such as BASF, Bayer and DuPont, which are positioned in many markets. Hybrids operate at all levels of the industry from upstream basic chemicals right down to specialties, agrochemicals, and even pharmaceuticals. They represent the largest level of participation, accounting for almost 40 percent of total sales.

Although it is large – generating USD 1.2 trillion in annual sales (excluding pharmaceuticals) at a conservative estimate or roughly 3 percent of worldwide gross domestic product – the chemical industry has only grown slowly in recent years. The recent wave of mergers and acquisitions (such as Degussa-Hüls and

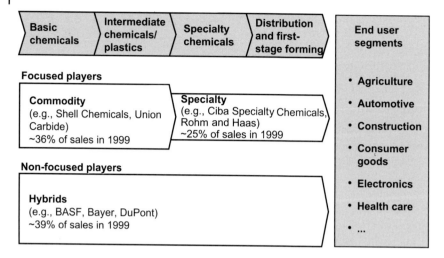

Fig. 3.1 Hybrids dominate the chemical industry today
Source: McKinsey

SKW Trostberg, or TotalFina and Elf Aquitaine, or the Dow Chemical/Union Carbide deal) has only changed the pattern slightly and the same companies have held the top ranks in the industry for well over ten years.

New players have not so far proved a real driver of change in this mature market either, except in certain segments. Notable newcomers are Asian companies, such as some Korean players, new startups such as Symyx, financial players such as Advent International and the Sterling Group and Internet-based exchanges such as ChemConnect. Clearly, many of the new entrants are not chemical producers, but knowledge-based or professional companies with distinctive skills to offer to the industry. However, the overall impact on the industry of the New Economy has been relatively small, with New Economy players and other new entrants accounting for less than 5 percent of total global chemical sales.

Firms are still in relatively good shape, however, because over the past fifteen years or so the upheavals in the industry have led to major restructuring, as the North American example shows. In the 1970s, huge value was created, mainly through double-digit growth rates, but much of it was destroyed in the early 1980s. Through the major restructuring wave of the late 1980s and early 1990s, the chemical industry got back onto a profitable track, and has not put in a bad performance in recent years (Fig. 3.2).

This development resulted in a relatively satisfactory stock performance in comparison with other basic materials industries such as steel and, in general, was more or less in line with the market as a whole until the mid 1990s (Fig. 3.3). The Asian crisis, however, proved quite a setback – wiping an estimated USD 25 billion off the industry's economic profit (return on invested capital minus WACC times invested capital) for Asian companies alone. The recovery of the Asian markets saw no matching improvement in the perception of the chemical industry by

Fig. 3.2 Improved profitability compensates for low growth in North America
Source: Datastream, McKinsey analysis

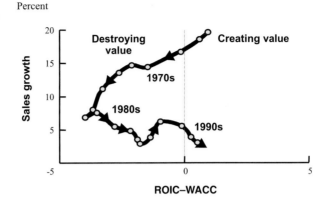

TRS (Total Return to Shareholders) index 1990=100

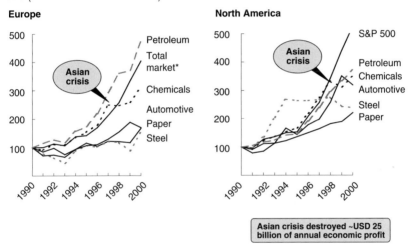

Fig. 3.3 Chemicals sector as profitable as total markets
* European Datastream index; Source: Datastream, McKinsey analysis

its shareholders, because the advent of the New Economy had made investors more skeptical about the growth prospects of the "old" one, and had in any event provided a strong counter-attraction. The chemical sector is now faced with two questions: first, how can it regain its pre-crisis profit level? Second, what forces will drive developments in the industry and help it to regain the approval of its investors? It seems to us that the recently developed stock-market-based performance metrics are both a strong driver and an excellent orientation point for chemical companies (see Chapter 2). Using these as the guiding principle and facilitated by new technology developments in both products (e.g., biotechnology) and delivery (especially through e-commerce and e-communications generally), chemical companies will be able to step up performance on the basis of tailored combinations of more efficient traditional asset-leveraging strategies and innovative and imaginative knowledge-leveraging ones.

Overall, the highly competitive and transparent environment with strong performance pressure will force companies to change their ways. New skills will become more important. Companies will have to move away from being "pure chemists" to expand both their financial and business competence, imposing a more rigorous and professional approach to their organizations rather than focusing solely on the product. For many, there are still value creation opportunities in improved marketing of their products, since historically marketing has not been a forte of the chemical industry. Instead of making molecules, they will have to become real service providers and generate solutions to their customers' problems, often beyond the supply of chemicals.

3.2
Strategies for Value Creation

The classic means of generating value through building and maximizing the return on fixed assets will remain important in this industry, but seems likely to decrease in importance over time. Organic growth of the business resulting from increasing demand is already priced into the shares. This also holds true for cost reduction efforts, in particular if they only compensate for the ever-falling prices of most chemical products. Unfortunately, these industry-wide cost savings also result in flattening industry cost curves, and thus increasing competition. In addition, the opportunities to generate more economies of scale through further industry consolidation are on the point of exhaustion for many product segments in North America and Europe, because supply side concentration is already very high.

Opportunities for portfolio restructuring, consolidation (especially in Asia), and the reduction of cost to core cost levels still offer value creation opportunities for many companies. Step function type increases in their EBITDA margin due to improvements in their functional skills, in particular in marketing and operations, will remain a major source of value. Importantly, however, we are talking here about an enhanced model: we are seeing firms using asset-leveraging strategies in a highly targeted manner to focus their business on their key strengths.

Ultimately, however, we believe that long-term growth can only be guaranteed by developing additional new skills and cultivating innovation to find new ways of value creation that are recognized by stock markets. As a result, companies will need to supplement the classical asset-leveraging strategies by a second element, knowledge-leveraging strategies.

Knowledge is becoming an increasingly important part of every company's capital, and so far it has not been captured sufficiently. There are several ways in which the chemical industry can create value through its use of knowledge –

many of which are strongly correlated to developments in the New Economy. We expect this to be the main source of future profitable growth.

A rough estimate of how the chemical industry will develop in the near future reveals that, over the next decade, asset-leveraging changes will still be the "bread and butter" strategy. However, as mentioned above, we expect to see a parallel here with other mature industries, such as the automotive sector, where all the cost-cutting and consolidation in the world has not done a great deal to move share prices. Companies do need to create significant additional profits to maintain the level of profitability of 10 years ago in the future as well.

3.2.1
The Focus of Asset-Leveraging Strategies

Let us look first at asset-leveraging strategies based on corporate restructuring and consolidation (Fig. 3.4). The hybrid chemical companies have begun to disaggregate as the opportunities for capturing synergies among different businesses dwindle or even die. They are starting to focus on those businesses where they have a leading position and are divesting the others. Key examples in the US market are Monsanto, which spun off its chemicals division in 1997 to form Solutia in order to focus on pharmaceuticals and agrochemicals and subsequently merged with Pharmacia & Upjohn to form Pharmacia, and Hercules, which divested its aerospace business and strengthened its water treatment business by acquiring BetzDearborn.

In the European chemical industry, there have been two main approaches over the past few years: "radical" redesign and "smoother" corporate development.

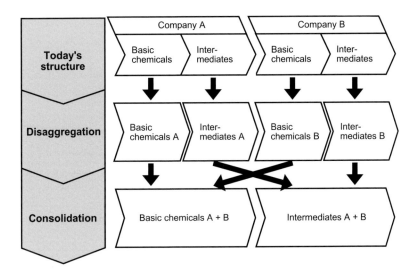

Fig. 3.4 Hybrids will transform into focused players
Source: McKinsey

Companies that choose the former approach radically refocus their entire business portfolios via extensive mergers and acquisitions, like Hoechst or ICI. Those selecting the smoother transition, such as BASF, target their efforts at strengthening existing core businesses through operational excellence and selected M&As.

Hoechst transformed its business from a hybrid chemical company into a pure life science company by radically divesting over 80 percent of its portfolio, getting rid of businesses such as specialty chemicals (sold to Clariant) or chemical intermediates (spun off into Celanese). It completed the acquisition of Roussel-Uclaf and bought Marion Merrell Dow, and finally merged with the French life science company Rhône-Poulenc after the latter had also spun off all its chemical business as Rhodia. By "reinventing" itself under the brand name Aventis, it increased its Return on Sales (ROS) from 4.6 percent to 12.4 percent, a development that has recently been viewed very favorably by capital markets.

BASF chose the smoother approach during the 1990s. It strengthened existing core businesses through selected divestment of assets and subsequent mergers and acquisitions. It began by getting out of various businesses such as chemical trading, magnetic tapes and polymethyl methacrylate and improving its operational performance in the other segments. At the same time, BASF acquired new businesses in its core areas such as acrylonitrile-butadiene-styrene (through the acquisition of DSM's ABS business) and crop protection (through the acquisition of the agrochemicals business of American Home Products) and showed organic growth in the businesses it kept. The result was an increase in its Return on Sales from 2.6 percent in 1993 to 10.0 percent in 1998, yet the overall shape of BASF did not change dramatically compared to the Hoechst case above.

To successfully transform major companies like these, two conditions have to be met. First, such firms need to have very strong leadership with a clear vision, combined with real commitment to the restructuring effort. Second, the models

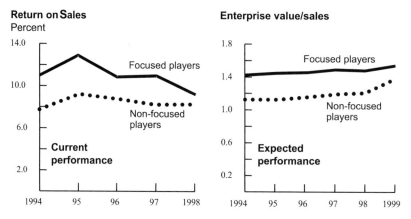

Fig. 3.5 Compared to hybrids, focused players performed better
Source: McKinsey analysis

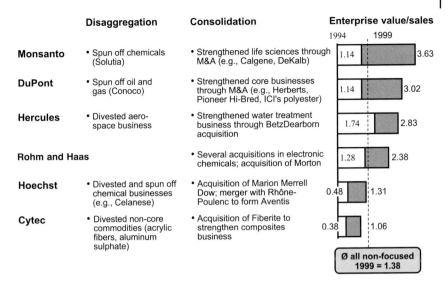

	Disaggregation	**Consolidation**	**Enterprise value/sales** 1994 ‧ 1999
Monsanto	• Spun off chemicals (Solutia)	• Strengthened life sciences through M&A (e.g., Calgene, DeKalb)	1.14 → 3.63
DuPont	• Spun off oil and gas (Conoco)	• Strengthened core businesses through M&A (e.g., Herberts, Pioneer Hi-Bred, ICI's polyester)	1.14 → 3.02
Hercules	• Divested aero-space business	• Strengthened water treatment business through BetzDearborn acquisition	1.74 → 2.83
Rohm and Haas		• Several acquisitions in electronic chemicals; acquisition of Morton	1.28 → 2.38
Hoechst	• Divested and spun off chemical businesses (e.g., Celanese)	• Acquisition of Marion Merrell Dow; merger with Rhône-Poulenc to form Aventis	0.48 → 1.31
Cytec	• Divested non-core commodities (acrylic fibers, aluminum sulphate)	• Acquisition of Fiberite to strengthen composites business	0.38 → 1.06

Ø all non-focused 1999 = 1.38

Fig. 3.6 Hybrids have started the transformation
Source: Datastream, annual reports, McKinsey analysis

work best with a high-speed approach. Hoechst, for example, managed to totally redesign and sell over ten businesses in the space of just two years.

Whether companies go for the radical or the smooth approach, empirical evidence shows that asset-leveraging strategies will pay off if their aim is to focus the portfolio. Focused players consistently show better performance over longer periods than non-focused players do, in terms of both return on sales and enterprise value (Fig. 3.5).

For example, Cytec divested its non-core commodities, such as acrylic fibers, acquired Fiberite to strengthen its position in the composites market, and increased its enterprise value to sales ratio from 0.4 to 1.1. Figure 3.6 shows some companies that have successfully focused on their core businesses by disaggregation and consolidation of their portfolios. This process is not yet finished, due to the high degree of fragmentation in some segments and the large number of hybrids. Therefore, the potential for value creation is immense and the opportunities for further disaggregation and consolidation are far from exhausted, though they will ultimately come up against natural limits.

3.2.2
The Ideas behind Knowledge-Leveraging Strategies

The search for new molecules has not been a source of significant growth since the late 1960s. Since the invention and commercialization of linear low-density polyethylene in 1973, for example, there have been no major new polymers in the chemical industry, and this situation is unlikely to change in the next five to ten years. While biotechnology may revolutionize certain fields, this remains a long

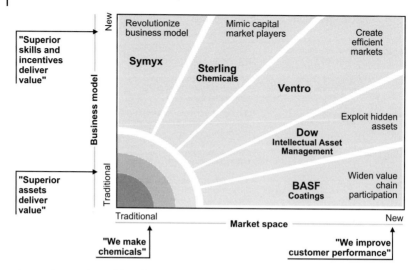

Fig. 3.7 Non-traditional knowledge-leveraging growth strategies
Source: McKinsey analysis

term prospect for many segments. In the meantime, there is little chance of new molecular innovations generating the kind of growth that will satisfy shareholder expectations either through new applications of existing chemicals or the substitution of chemical products for natural materials.

Our set of knowledge-leveraging strategies is based on the exploitation of knowledge. This knowledge can be developed and applied along two dimensions: market space and business models (Fig. 3.7). Along the market space dimension, companies will change from the traditional mindset of making molecules to focusing on new customers, who are, typically, downstream. Along the business model dimension, the industry has traditionally followed the maxim that superior assets deliver value – and they actually do. However, it is becoming ever clearer that "intangible assets" such as superior skills and incentives are additionally needed to bring the performance of the chemical industry back to its pre-Asian crisis levels.

There are at least five knowledge-leveraging approaches that companies or business units can explore. All five entail a change in the way that chemical companies have traditionally thought about their businesses. Many will require much less capital than traditional strategies, but each can improve performance significantly.

Revolutionize the business model. Better ways can be found of doing the same things, the most obvious example being the use of information technology and the Internet to improve customer service or corporate organization. Process industries have been slow to take up these enabling technologies, perhaps because the broad implementation of ERP programs did not improve their individual performance to a degree that would have excited shareholders. Nonetheless, automation and electronic improvement of every step of the internal value chain can dramatically eliminate costs and create a step change in company profitability, though this will only result in value creation if an individual company clearly outperforms industry standards.

One of the most potentially lucrative approaches for value creation is the development of entirely new processes to make existing products. Over the next ten years, biotechnology and combinatorial chemistry are likely to change the face of quite a number of industry segments even outside of agrochemicals or food ingredients, as well as the way that chemical companies produce molecules.

An example of this approach is Archer Daniels Midland, which reduced total production costs by over 60 percent by replacing traditional chemical synthesis with an advanced biological fermentation process for the manufacture of L-lysine. It is now the dominant worldwide producer with a market share of 27 percent and a process that allows it to compete on cost with Asian producers.

Symyx has become a pioneer in the use of combinatorial chemistry and high-throughput screening as a method for discovering new materials for chemical and electronic applications. Its approach is up to 100 times faster than traditional research methods and reduces the cost per experiment to as little as 1 percent of that of traditional research methods.

Mimic capital market players. Those players currently reaping most profits from the chemical industry are not chemical companies but financial players – venture capitalists, private equity funds and principal investors – who are adept at spotting opportunities for value creation, and who excel at extracting that potential value, such as The Sterling Group, Advent International, Morgan Grenfell, and CVC (Citicorp Venture Capital).

For example, by instituting profit sharing, employee ownership, and stock option plans, The Sterling Group succeeded in encouraging managers at Cain Chemical to reduce overheads by up to 60 percent, increase operating margins by 7 percent and raise throughput by 25 percent. It then sold the company, reaping over USD 1.1 billion in profit from a USD 28 million investment (see also Chapter 8).

Venture capitalists use some of the same techniques. Their business model encourages success and, in the absence of success, leads them to quickly prune the investment. By providing the right incentives, venture capitalists push innovators to transform their ideas into lucrative businesses. Given the choice, surely most young innovators would prefer to be backed by a determined venture capitalist than a large chemical company with relatively sluggish R&D processes and modest compensation packages.

Most chemical players consider these managerial techniques the domain of Internet players, but financial players are using them successfully to rob the chemical industry of both value creation opportunities and talent. However, there is no real reason why chemical companies that possess strong managerial capabilities cannot identify and develop similar business opportunities, especially since they naturally have a deeper understanding of the chemical market than most outside investors.

There are several ways in which chemical companies can do this. First, they could apply the same kind of management tactics and incentives to their own businesses to reduce overhead costs and eliminate internal subsidies – a significant problem in traditional chemical companies, where a lack of cost/price transparency makes it difficult to identify weak businesses. Second, they could set up their own venture capital funds to encourage innovation and attract new talent. More ambitiously still, they could create their own funds to buy up companies in distress then use their own management skills to turn these companies around.

Create efficient markets. Despite the fact that many chemicals are considered to be commodities, much of the market is still bilateral, meaning most relationships are directly between the supplier and customer, with no intermediaries. This has often been in the chemicals companies' interests, as the lack of transparency allows them to maintain slightly more favorable margins. However, chemical companies have to accept that, in a world of e-commerce, more transparent pricing is inevitable, and some products – especially commodity products – will lend themselves to market trading.

For some chemical companies, this is a nightmare scenario that threatens to send prices tumbling by forcing them to compete on their product's intrinsic value – a value which commodity products typically lack. However, there are also substantial value creation opportunities in the establishment of more efficient markets, both for those that set up the marketplace, and for those who participate. While companies can take the short term view that efficient markets destroy their ability to increase margins through opaque pricing schemes, leading-edge chemical companies will recognize that these new markets create opportunities for intermediation, risk management, and commodity trading, just as financial products and their derivatives do for financial institutions today.

Indeed, a number of incumbent players have started to establish online marketplaces, by themselves or together with newcomers, following a model set up not by chemical companies but by Internet brokers such as Omnexus or Elemica (see also Chapter 7).

Eventually, transparent pricing and trading could lead to the introduction of financial derivatives for some major chemical products such as methanol, low-density polyethylene and styrene. If a forward price curve were established for such products, manufacturers would be better able to time new asset investments and lock in margins through price guarantees on inputs and outputs.

Such a market for trading chemical contracts would effectively help the industry separate the risk of asset ownership from both production and financial risk. Imagine, for example, if a pure commodity player could conclude forward contracts to lock in the price on most of its output volume and input raw materials. It could outsource the sales and logistics process to efficient transactional market makers and low cost logistics specialists, leaving it to focus purely on being distinctive in low cost operations.

Once such financial markets for chemical products exist, we will most likely see the emergence of chemical traders – similar to Enron in energy – that disaggregate and reaggregate risk both for themselves and others. Rather than making chemicals and selling them to generate value, companies that trade chemically backed financial derivatives will simply be using their knowledge of the market and the industry to reap profits. This is quite a leap forward in thinking for companies that have largely seen value creation in the building and operation of plant, and the production of chemicals.

Exploit hidden assets. This strategy entails applying available knowledge, assets, and skills in new ways. Over time, the chemical industry has built a broad base of non-physical assets such as brands, patents, customer information and institutional skills. Only a handful of companies, however, are trying to use these to their maximum economic potential.

DuPont is one of the few. It is renowned in the industry for its ability to run plants safely and it enjoys fewer lost working days from accidents than anyone else in the industry. Several years ago, it decided to exploit that institutional skill and started a consulting business specializing in plant safety. Another example is Dow Intellectual Assets, a subsidiary of Dow, which has been extracting value from licenses and patents by using a multidisciplinary team that examines patents for licensing opportunities. As a result, licensing revenues from patents have risen from USD 25 million to over USD 125 million.

Widen value chain participation. Finally, in some segments of the chemical industry (in particular specialties), opportunities still exist to use the traditional skills of R & D and application development, but to target them at new customers further down the value chain. This move mirrors the approach attempted by many in the industry twenty years ago when they shifted portfolios from commodity to specialty chemicals in search of higher margins. It was not a particularly successful strategy at the time, both because of the number of players that sought to become specialty manufacturers and because the capital markets were skeptical about the growth opportunities for specialties.

This time, however, if companies keep a much stronger focus on the needs of downstream customers and deliver real value, they might fare better. Rather than selling its rubber to producers of medical gloves, Dow, together with Maxxim Medical, has formed its own gloves business based on polyurethanes. Though the business is still developing, it serves as an example of how companies can begin looking for new growth opportunities by understanding the value of their own products and not letting this value accrue to downstream producers.

Similarly, BASF Coatings has exploited its advanced knowledge of paints and the willingness of car manufacturers to pay for painted cars to capture downstream value. They have deals with some of the leading automotive companies to actually paint their cars, rather than simply selling them paint. BASF recognized that car manufacturers did not need to own the painting process and, in fact, were happy to simply purchase the "finished product." Using its understanding of the coating process and chemistry, the paint company improved the quality of painted cars and captured higher margins from its business as it used its own knowledge to reduce the consumption of paint.

3.3

The Impact of Value Creation Strategies

As we see it, the move toward more focused product strategies and new approaches to leveraging knowledge will affect the industry landscape of the future in four main ways (Fig. 3.8).

First, the process of building new activities along the value chain will bring the chemical industry even further toward the end customer as companies reach out to new groups of customers further downstream. At the same time, we will see an even further increase in the commoditization of specialties because customers will grow more sophisticated, and producers will not be bringing anything new to the table.

Second, the complex hybrids containing chemical and non-chemical businesses under one roof will continue to break up and spin off non-chemical businesses. Conglomerates with chemical and non-chemical activities will disappear or at least decline significantly. The empirical evidence is almost overwhelming that the corporate centers of these conglomerates find it very difficult to add value. Synergies between chemical and other businesses seem to be either non-existent or difficult to capture. The resulting "conglomerate discount" creates a strong incentive to de-merge these companies. Some transactions that already have taken place are the IPOs of Conoco (DuPont) and Agfa (Bayer). BASF sold its chemical trading and magnetic tapes interests, and Dow Chemical sold its pharmaceutical business Marion Merrell Dow to Hoechst. But there is still a lot to do.

Third, the industry will start to consolidate to form high performing "global slivers," companies which have captured global leadership in a single business or

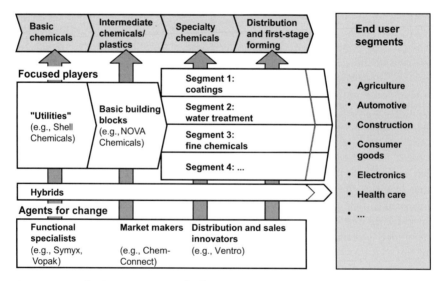

Fig. 3.8 Potential industry structure in the future
Source: McKinsey

segment on the commodities and the specialties side. On the commodities side, we will see the classic oil companies, for example, owning more utility-type activities, and potentially some focused polymer companies such as Basell. On the specialties side, focused players will start to form around particular segments such as coatings or water treatment.

Finally, more new players will start to enter the market. These may be either incumbents experimenting with new business models and new market spaces such as DuPont, or there may be new players. We basically see three types of new entrants emerging: functional specialists (i.e., infrastructure providers such as Vopak), market brokers such as CheMatch or Omnexus, or financial investors such as CVC and The Sterling Group.

If there were no constraints, this logic would eventually entail each company focusing only on one chemical segment, resulting in many highly specialized small and medium-sized companies. But, although focus is important for profitability, there are limitations to the disaggregation of chemical companies. Companies need sufficient market capitalization for critical mass to protect them from takeovers and to secure their attractiveness in the capital as well as the recruiting market. We estimate that, for the chemical industry, this market capitalization will be roughly USD 10 billion in the medium term, equivalent to the sales volume. Given the typical size of chemical markets, companies focused on only one segment will not make the grade. Therefore, in all likelihood we will see more complex structures. "Chemical conglomerates" will still exist and retain several business segments to maintain critical mass, and companies will concentrate on business segments where they can occupy the top or second market position.

Chemical companies are now faced with a choice. If they go on down the beaten track of the last decade, their share prices will in all likelihood follow the trend of other mature industries. But if the industry can get out of its comfort zone and really step up to the challenge of leveraging existing knowledge and developing new skills, new prosperity may be just around the corner.

4
Managing Commodity Portfolios
Thomas Röthel, Gary A. Farha, and David F. Hoffmeister

At a time when other industries are experiencing dynamic and unprecedented change, with major portfolio upheavals and broad company repositionings, many chemical companies have remained integrated, diversified conglomerates. To many observers, they seem stuck in a bygone era. However, the industry has started to show signs of becoming more focused and entrepreneurial, and we feel that this trend has a lot further to go. To play this game successfully, chemical companies will have to free themselves of traditional vertical and horizontal structures and have the courage to own only those parts of businesses where they can add excellent value.

4.1
The Shakeup has Begun

There are some exciting changes taking place in the chemical industry, and it seems that much more is yet to come. New structures and increasing competition are beginning to reward those companies that pursue more focused and more entrepreneurial strategies. Like steel, pulp and paper, and petroleum before it, the chemical industry is fast finding that the old model – of complete ownership of a vertical industry chain plus diversification across numerous product segments – is no longer the most successful way to compete.

Historically, the major players in the industry have been jacks-of-all-trades. Integrated businesses in which a single company controlled the entire process – from the supply of raw materials to the outlets for its products – were the norm. Because of the development of individual national markets and the lack of a significant pipeline infrastructure, Europe became even more integrated than North America.

The diversity and complexity deepened during the 1970s and 1980s as many companies moved downstream into specialty chemicals and other areas. Such an approach was essential for them to remain cost competitive. It allowed them to capture the benefits of physical integration – namely, a secure supply of feedstocks, convenient outlets for intermediate products, minimal logistics costs, and the optimal use of byproducts. Integration also helped to reduce problems in an environment where many raw material sources and product markets were controlled by a small number of players.

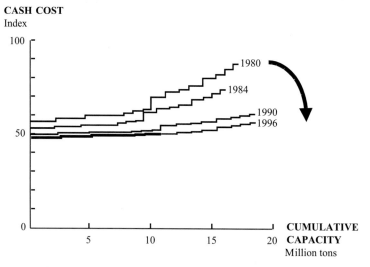

CASH COST
Index

Fig. 4.1 Flattening cost curves – development of ethylene, USA
Source: TECNON, McKinsey analysis

Number of producers*

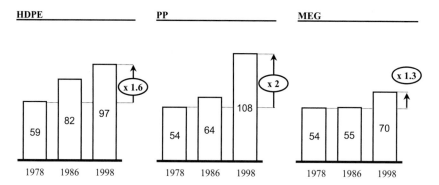

Fig. 4.2 Number of producers is increasing worldwide – examples: HDPE, PP, and ethylene glycol
* Joint ventures counted as one player; Source: TECNON, McKinsey analysis

Today's chemical industry is more competitive and mature. But it faces several new challenges that call its traditional structure into question:

- Growth in the industry, which was several times that of GDP in the 1960s and 1970s, has slowed considerably in the United States and Western Europe.
- Pressure on prices has developed for commodity chemicals as technology has diffused ever more rapidly, flattening industry cost curves and impeding differentiation (Fig. 4.1). An analysis of a basket of commodity petrochemicals reveals a real decline in gross margins of 2 percent per annum in the USA.

- A rising number of new competitors have entered the industry as a result of high growth rates in newly industrialized countries, the deregulation of chemical markets, and privatization (Fig. 4.2). The number of polypropylene producers worldwide, for instance, doubled between 1978 and 1998.
- Falling trade barriers, heavy investment in handling facilities, and improvements in transport have boosted the increasingly liquid markets for commodity and intermediate products. Between 1981 and 1995, cross-border trade volumes grew at double the growth rate of the industry as a whole. Factors like these are undermining the historical benefits of owning assets. Why make upstream products when you can buy them elsewhere more cheaply?

The forces now buffeting chemical companies are uncomfortably reminiscent of the changes that brought about the demise of "Big Steel" in the 1970s and 1980s. Large integrated producers in that industry were devastated by intensifying competition from imports and a host of "mini-mills" that adopted new technology which was able to undercut the cost structure of the older furnaces.

This is not to say that a similar bloodletting is in store for the chemical industry. For one thing, none of the new players seems to have as big an advantage in technology and cost as the mini-mills enjoyed in steel. Moreover, many big chemical companies have responded to the structural changes and fiercer competition of the 1990s. There have been a variety of spin-offs and sales of large divisions, organizational restructuring programs, and improvements in efficiency. All have been helpful, but few have gone far enough to address the deeper challenges facing the industry.

Efforts at disaggregation a few years ago (like ICI's demerger of its pharmaceutical division, Zeneca, and Union Carbide's spin-off of its industrial gases division, Praxair) have had a big impact on the value of the spun-off entity. However, spin-offs of non-core divisions do not automatically improve the retained chemical businesses, in part because what remains may still be too diversified.

However, players such as Crompton and Knowles, NOVA Chemicals, Great Lakes Chemical, Huntsman, Methanex, Shell Chemicals, Georgia Gulf, and companies owned by financial investors (e.g., Advent (Vinnolit) or G. Harris (Vestolit)) are now leading the way in restructuring and in developing new skills. Between 1985 and 1995, Great Lakes and Crompton, for example, generated annual returns to shareholders of 23 percent and 26 percent respectively, while the index for all chemical companies stood at just 9 percent. Huntsman grew from sales of just USD 200 million in 1983 to over USD 4.3 billion in 1995, while Methanex increased its share of the methanol market from 21 percent to 38 percent between 1993 and 1996, becoming the global leader in the process.

These companies have created core capabilities in areas other than production. They have, for example, used superior deal-making and human resources skills to expand rapidly. Moreover, they have focused their businesses by acquiring and merging business units from less focused competitors. With these, they have built up powerful positions in product innovation or cost leadership – like that of Nova, for example, in polystyrene.

Many of the old integrated chemical companies have belatedly begun to make significant changes. ICI has taken steps to reshape itself through its acquisition of Unilever's specialty chemicals businesses and the subsequent disposal of its commodity chemicals. During the nineties UCC transformed itself into a pure petrochemicals player. Similarly, NOVA Chemicals has focused its business on olefines/polyolefines and polystyrene. Over the last few years it has separated out the pipeline business and reduced its stake in Dynegy. DuPont has become more focused on the performance of its individual business units while Shell Chemicals has focused its petrochemicals portfolio radically over the past 24 months.

4.2
The New Game – Disaggregation and Focus

We have been observing this trend in the industry and, from conversations with chemical executives, we sense that the industry is poised for greater changes. We believe that nothing less than a new industry structure is in the making – one in which companies compete only in those product segments where they can aspire to lead.

The result of this will be a stronger industry. Size will still be an advantage, but less so in vertical markets than in horizontal ones. Companies that accomplish the transition from vertical integration to disaggregation, and from being diversified to being focused, stand to create considerable shareholder wealth. Even more value will be captured if they adopt new approaches to organization and management – for example, by focusing on shareholder value and by developing expertise in acquisitions and deal-making.

In other words, what integrated and diversified chemical companies need now is bold action aimed at turning them into focused entrepreneurial organizations. Those seeking to make such a transition should first focus on key product or market segments, and then consolidate at the business unit level or below. Once companies have taken these steps, they should use their new platform as the basis for an ambitious growth plan. If they are to execute each step successfully, however, companies will also need to create a more entrepreneurial culture. The result will be an industry that looks very different from today's.

The first step in unlocking latent value in the industry is to direct effort and investment more toward selected product or market segments. An analysis of the performance of American chemical companies over the last industry cycle suggests that those competing in a focused fashion tend to do better than those competing more broadly. In a large sample, focused companies had a 30 percent higher average return on invested capital than unfocused ones. Among the top performers, the difference was a full 90 percent.

The advantages of increased focus are apparent in other process-intensive industries. In pulp and paper, for example, there is a strong correlation between a company's degree of focus and its financial performance. None of the industry's top performers (as measured by shareholder returns) competes in more than five different segments.

Focused companies like these perform better because:

- Managers have a better understanding of the key success factors for individual businesses, and this makes them less likely to stretch the same skills inappropriately across different areas.
- Cross-subsidies are eliminated and uncompetitive businesses are divested or left to "go bankrupt".
- Capital and management talent are allocated more efficiently.
- The overheads needed to manage a diverse set of operations are reduced and can add more value due to the similarity of the businesses.

Most chemical companies today are organized in large divisions that, in turn, consist of many different businesses. The smaller units tend to be integrated with other businesses in their divisions rather than standing alone. Even where separate business units do exist, their managers often control less than half of their P&L. Manufacturing and the purchase of raw materials and services are never entirely under their jurisdiction. To make matters worse, they are usually burdened with overheads from other parts of the company. This lack of real managerial accountability, coupled with squabbling between units over non-market-based transfer pricing, often leads to second best decisions.

In order to make a focused strategy work, chemical companies must break up their integrated, vertically-oriented structures. This should enable them to highlight the advantages and disadvantages of competing in various segments. Disaggregation may, for instance, highlight a production advantage that was previously masked by transfer pricing. Across the industry as a whole, it may lead to improved pricing. Prices could be set independently by the two highest marginal cost producers rather than by an integrated producer using a dubious "rolled cost" calculation (i.e., integrated marginal contribution and margin-oriented calculation).

Several companies have completed or are in the process of completing this first step, among them the company formerly known as Hoechst (now Aventis and Celanese), Dow, DuPont, Shell Chemicals, NOVA Chemicals, ICI and Degussa-Hüls. However, not all of them have taken the reorganization down to the business unit level – a necessary condition if they are to reap the full benefits.

The next trick is to identify those segments that can best drive success, and then to exit from all the rest. This requires a company to analyze its entire portfolio, assessing performance, deciding which businesses are likely to be attractive in the future, and selecting those in which it can be a winner. It must map out the capabilities needed in all the product segments in which it competes, and then it must determine where its own core capabilities are strongest.

Once the core businesses and capability platforms have been identified, the remaining businesses should be divested through sales, spin-offs, or joint ventures. If this is done well, exiting from selected businesses can simultaneously bring more focus to the survivors. The core units should then be grown by consolidation and organic growth based on new products, geographic expansion, or a revamped value delivery system. To be successful, these changes need to be supported by an entrepreneurial organization with small connected teams, significant incentives, and sufficient freedom for the management.

There are no hard and fast rules limiting the number of businesses that a company can be in. The only constraints are that it should be distinctive in each of them, and that its portfolio should not incorporate too many dissimilar businesses.

Focus alone does not guarantee success, however. While focused chemical companies perform better on average, individual company returns show wide variations. Picking the wrong businesses on which to focus makes it difficult to create value, no matter how good the execution. Conversely, weak execution even in the right businesses leads to poor financial performance.

4.3
Overcoming Barriers to Action

The prevailing wisdom is that many barriers stand in the way of chemical companies when they try to create the kind of focused organization that can capture the industry's value potential. Barriers certainly exist, but they are lower than many believe, and they are crumbling. Creative solutions can dismantle them quite painlessly.

Consider three common beliefs that prevent chemical company executives from taking action:

"We've got to be vertically integrated". It is often thought that companies have to remain vertically integrated if they are to be efficient and cost-competitive, and that they therefore cannot afford to focus. This barrier can be overcome if a company abandons its organizational integration, takes advantage of the growing merchant market for chemical products, and uses creative "quasi-integration" options.

Physical integration remains necessary in some cases and beneficial in others. In general, though, companies do not need to be *organizationally* integrated or to own each step of the product chain to be physically integrated. By confining integration to the physical level, companies can avoid such drawbacks as cross-subsidization, lack of market transparency, and poor investment and pricing decisions. At the same time they can retain the benefits of lower logistics costs, shared services, and the utilization of byproducts.

Many companies resist this logic, arguing that they have to remain vertically integrated because they have no other way of purchasing feedstocks or of selling intermediate products. In reality, though, growing merchant markets exist for most products. Even with such a major raw material as ethylene, numerous companies – among them Borealis, Celanese, Degussa-Hüls, DSM, Georgia Gulf, and Solvay – rely on the merchant market for some or all of their facilities. What is more, players at some integrated sites have also switched to the market. Arco is now selling styrene on the merchant market, for instance, and Bayer is purchasing the raw materials for ABS and NOVA those for PS in Europe, instead of making them.

The rapid development of "colocation sites" around the world is enabling ever more companies to execute this concept of disaggregation and focus (Fig. 4.3). To-

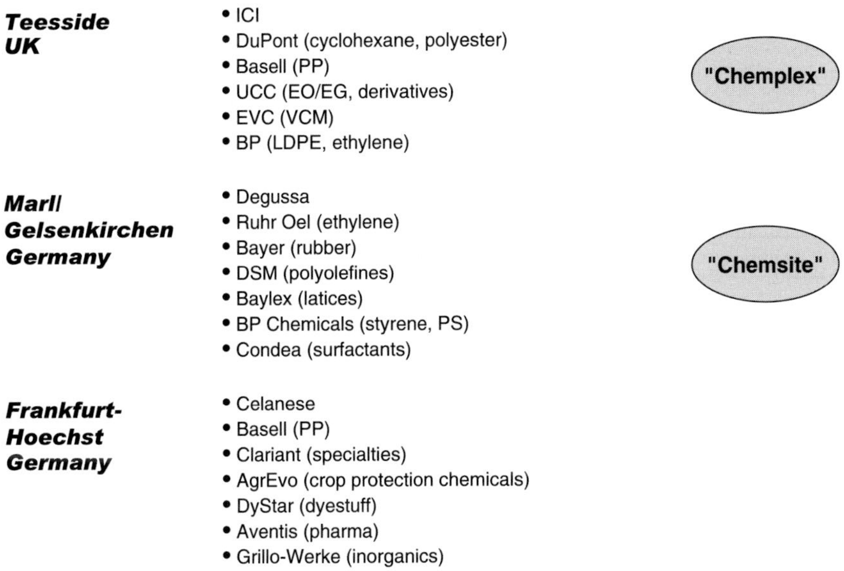

Teesside
UK

- ICI
- DuPont (cyclohexane, polyester)
- Basell (PP)
- UCC (EO/EG, derivatives)
- EVC (VCM)
- BP (LDPE, ethylene)

"Chemplex"

Marl/
Gelsenkirchen
Germany

- Degussa
- Ruhr Oel (ethylene)
- Bayer (rubber)
- DSM (polyolefines)
- Baylex (latices)
- BP Chemicals (styrene, PS)
- Condea (surfactants)

"Chemsite"

Frankfurt-
Hoechst
Germany

- Celanese
- Basell (PP)
- Clariant (specialties)
- AgrEvo (crop protection chemicals)
- DyStar (dyestuff)
- Aventis (pharma)
- Grillo-Werke (inorganics)

Fig. 4.3 Colocations are becoming more and more common
Source: McKinsey

day, sites like Teesside (UK), Marl/Gelsenkirchen, Frankfurt-Hoechst (both Germany), Tarragona (Spain), ARA (Belgium/Netherlands), several sites along the US Gulf Coast, Brazilian "polos" (e.g., Camacari), or Map-ta-Phut (Asia) offer non-integrated companies secure feedstock supply at competitive costs, together with world-class infrastructure and services. We believe that these and similar sites represent the concept for the future. The companies will individually concentrate on those processes they can deliver best, and thereby improve the competitiveness of the entire system (Fig. 4.4).

If a company is still concerned about ensuring the security of its material supplies or establishing an outlet for its products, a variety of solutions exist. It can guarantee feedstocks by getting a superior operator of the upstream business to commit to a long term supply contract. Shell and Borden demonstrated this kind of integration when they formed a "virtual cracker" in which Borden paid for the de bottlenecking of Shell's ethylene production facility to gain a cheap, secure feedstock supply.

Another way is to establish a "condo" cracker that is owned by a number of parties interested in its output (see Chapter 14). Some companies have devised other approaches – such as joint ventures and "host and tenant" structures – that allow them to turn over the management of certain businesses to their natural operators whilst maintaining a secure supply or outlet. Pursuing these quasi-integration options allows companies to enjoy the benefits of physical integration while capturing the advantages of focusing on selected core segments.

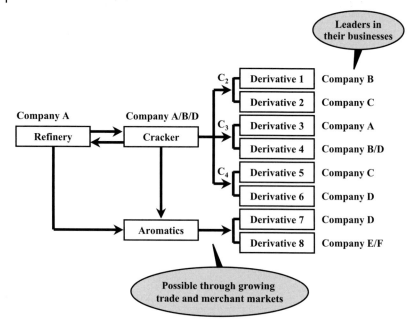

Fig. 4.4 Vision of future petrochemical complexes
Source: McKinsey

"Restructuring costs too much". Some believe that the restructuring costs involved in consolidation are so high that they outweigh the benefits. One frequently quoted example is the exit cost of closing plants. Many maintain that plant closures are essential if firms are to reap the benefits of improved structure and capacity rationalization, but that they are extremely costly or almost impossible to implement. Some firms have, however, found ways to reduce exit costs by staging partial closures or by finding alternative uses for plants.

In reality, plant closures account for only a fraction of the benefits of consolidation. In two recent post-merger cases in the chemical industry, none of the synergy value captured was attributable to outright closures. There are ways of rationalizing capacity and taking advantage of an improved industry structure without jettisoning existing capacity.

One option is to close just part of a plant in order to reduce capacity. A European chemical company shut down 20 percent of an ethylene derivative plant, cutting the plant's fixed costs by 20 percent and its personnel by 15 percent. The company was then able to find a merchant buyer for the ethylene that was produced. Despite the closing and severance costs that the company incurred and the effect on its ethylene business, the partial closure raised the facility's net present value by 35 percent.

"We will have to shrink the business". The final belief that ties the hands of chemical executives has to do with traditional management notions about the im-

portance of building assets. Believing that they will have to shrink their own businesses to achieve focus, many executives may be reluctant to play the new game. Certainly, companies usually need to divest some of their businesses before they can start to consolidate and grow. However, to avoid shrinking their overall business excessively, they should try to focus and consolidate virtually simultaneously, perhaps by using asset swaps.

Ultimately, though, companies must not let psychological barriers get in the way of the new approach. Value creation, not size, is what matters to shareholders. Companies that realize this and accept some temporary business shrinkage to play the winning game of the future will create tremendous value.

Growing pressure from shareholders for greater transparency and value creation, new entrants from Asia and the Middle East, and the emergence of a host of more focused competitors indicate that the time is right for big chemical companies to decide whether they are going to be among the first movers in the winning game or risk being left out. Chemical executives need to assess where their company stands and what it must do to succeed. They can begin the diagnosis by asking themselves the following questions:

- Are genuinely market-based business units in place? If so, what percentage of their costs and assets do the business unit managers control?
- What capabilities are critical to success in each business unit, and how distinctive are we in each compared to competitors?
- Do we explicitly compare our BUs' value to us with their value to others? Which businesses would we keep if we were starting a company from scratch? What are we doing about the others?
- What are we doing to improve the industry structure of the businesses on which we have decided to focus?

5

How to Succeed in the Rapidly Maturing
Specialty Chemicals Industry

Joël Claret, Simon Lowth, and David McVeigh

Historically, specialty chemicals have been among the highest growth and highest margin segments of the chemical industry. However, growth, margins, and performance have all recently declined as the segment matures rapidly. As a result, the share prices of specialty chemical companies have been anemic over the past five years compared to indices for either the chemical or the total market.

Specialty chemical companies, traditionally in a position to make relatively easy profits, are faced today with highly complex maturing portfolios which present them with limited options for growth. Furthermore, since market observers have seen new CEOs arriving at around fifteen specialty chemical companies over the last two years, it is clear that the industry is now desperately seeking new ways of achieving higher performance. To create and capture value in an extremely difficult market, these companies need to adopt a comprehensive and holistic approach to improving performance.

First and foremost, they should aim to outperform competitors by reshaping their portfolios in selected arenas (thus driving industry restructuring), optimizing the business model for each segment and executing it outstandingly, seeking organic growth, and putting more value-driven management in place at the corporate center.

5.1
The Commoditization of Specialty Chemicals – What Happens Next?

Until the mid-1990s, specialty chemicals were among the most attractive segments of the global chemical industry. Market growth in many specialties was high compared with that in commodities, often two to three times GDP growth. Gross margins and operating margins were also excellent, and most segments were shielded from the vicious cycles that plagued the commodities markets.

As a result, many chemical companies embarked on major corporate transformations targeted at shifting away from commodities and into specialties. A classic example was ICI. During the 1990s, it sold most of its commodity businesses and made huge purchases of specialty-focused companies, including National Starch and Chemical.

% CAGR, 1997 - 2002*

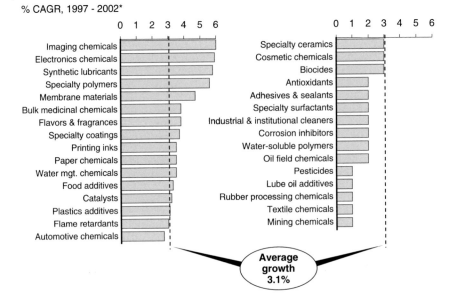

Fig. 5.1 Expected specialty chemicals growth in developed economies
* North America, Europe and Japan; Source: SRI, McKinsey analysis

At the dawn of the new millennium, however, the specialty chemicals business is not looking so special anymore. Volume growth in many segments has slowed to near GDP levels as markets mature and inter-material substitution rates slow (Fig. 5.1). The only exceptions to this appear to be in highly technical segments where the end-use market is growing rapidly (for example, electronics chemicals and imaging chemicals) and in geographic markets where the per capita consumption of chemicals is still quite low (in South America and South East Asia, for instance).

Prices and margins in many segments are also coming under increasing pressure. In the 1990s, chemical buyers improved their purchasing processes and skills. This led to more frequent competitive bidding, especially for semi-specialties where suppliers and functionally equivalent products abound. Furthermore, the customer base for many specialties is rapidly concentrating in segments such as automotive, personal care and food, significantly increasing customers' purchasing clout. Finally, increased competitive intensity is adding to price pressures as specialty chemical companies continue to expand their global reach.

This trend is likely to continue (and possibly accelerate) with the emergence of e-commerce and its market-making models such as exchanges and reverse auctions. The only areas that appear safe are those true specialties where suppliers have a unique technology or product.

Specialty companies have started to respond to these threats in two ways. One is by continuously reducing costs – including reductions in core specialty chemical functions like technical services and R&D. Many observers feel that these

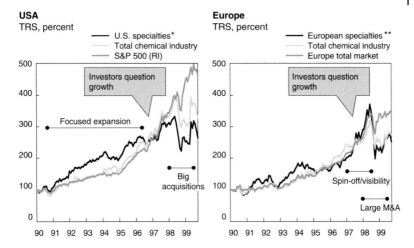

Fig. 5.2 Performance of specialty chemical companies
* Crompton, Great Lakes, Hercules, Morton Int., Rohm and Haas, Witco, Albemarle, Cabot, Cytec, Engelhard, Ethyl, Ferro, Fuller, Grace, IFF, Lubrizol, Nalco, Sigma-Aldrich;

** Ciba, Clariant, Degussa-Hüls, ICI, Allbright-Wilson (until 09/99), Allied Colloids (until 03/98), Courtaulds, Croda, Elementis, Laporte, Süd-Chemie, SKW Trostberg, Johnson Matthey; Source: Datastream, McKinsey analysis

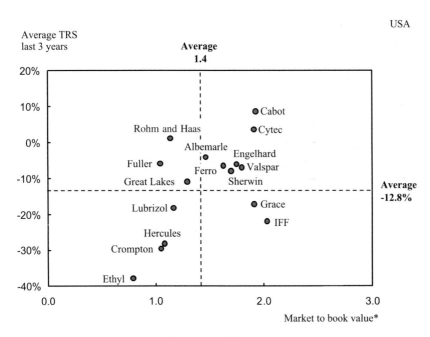

Fig. 5.3 Few players have outperformed the pack – USA
* As of September 30, 2000; Source: Datastream, annual reports, McKinsey analysis

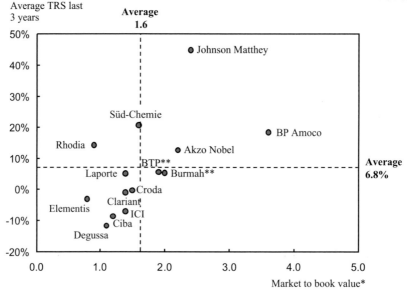

Fig. 5.4 Few players have outperformed the pack – Europe
* As of September 30, 2000; ** BTP acquired by Clariant (January 2000) and Burmah acquired by BP Amoco (March 2000); Source: Datastream, annual reports, McKinsey analysis

short term, tactical measures will help stem margin erosion, but that they may, in the long run, further reduce the growth potential of the industry.

The other common response has been mergers and acquisitions. The past five years have seen an unprecedented level of deals in which entire companies have been purchased in a drive for economies of scale (e.g., Rohm and Haas/Morton, Ciba/Allied Colloids). However, analysts believe that many of these deals were ill conceived, or that the purchaser paid too high a price, leading to a further deterioration in net incomes and market valuations.

As a result, the stock market performance of specialty chemical companies has lagged behind that of earlier periods. In fact, very few specialty chemical companies have done well over the last five years, as measured by their total return to shareholders or their market-to-book ratio, and very few have outperformed the pack in that period. This, in turn, has led to slower share price growth for the industry as a whole relative to broader market indices (Figs. 5.2–5.4).

For the specialty chemical industry and its players to regain the luster of the past and the confidence of the capital market, three changes are absolutely essential:

1. The industry has to be restructured according to arenas, requiring extensive portfolio reshuffling
2. New sources of growth have to be captured within and outside the current industry boundaries

3. A few shapers have to trigger change throughout the industry by successfully pursuing the right sequence of moves.

Will the industry landscape end up being consolidated by a few very large multi-business players (mega-specialty companies)? Or will it consist of a number of smaller players who are extremely focused on a few segments each? Or, for that matter, will new players enter the industry? It seems to us that its ultimate fate will depend on the ability of the few current largest specialty companies to lead the way and take a new approach to value creation. The purpose of this chapter is not to propose a likely endgame across the industry or even across individual segments, given the extremely wide diversity of specialty segments. Rather, we attempt to describe ways for individual companies to seek a holistic management approach, which – even more importantly than in other industries – has to achieve breakthrough performance improvement across most of their businesses and reshape the entire industry.

We believe that to achieve and sustain the essential performance improvement, specialty chemical companies should focus on four major themes:

- Actively recasting and optimizing their business portfolio to shape and achieve leadership positions in selected arenas. They can do this in a number of ways: by making carefully selected acquisitions and divestments, by consolidating fragmented segments of the industry, and by seeking new kinds of synergies between businesses.
- Refreshing and better aligning the business model pursued at the business unit level according to customers' needs and the stage of the segment in the life cycle, and then executing that model outstandingly.
- Pursuing new avenues for organic growth, both traditional ones such as product-based innovation and non-traditional ones such as new technologies, e-business and – perhaps the most exciting route – the building of service-based businesses.
- Redesigning the corporate organization in a way that best unleashes performance at business unit level and captures cross-business synergy, in particular putting in place a lean "investor-type" corporate center that can unlock these synergies at low cost.

5.2
Recasting the Portfolio

Recasting the portfolio is the first key lever that specialty chemical companies should pull to create sustainable value. It is particularly appropriate for those companies aspiring to play in the multi-billion dollar specialty chemicals "league" and to own a broad range of businesses.

A closer look at the various specialty chemical segments suggests that, for several reasons, most of them will undergo significant restructuring over the next decade.

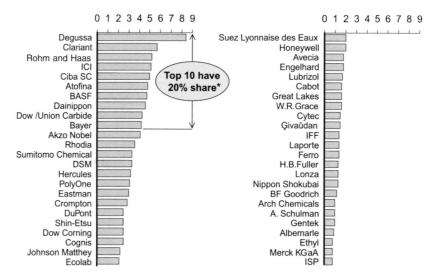

USD billions   ESTIMATES

Fig. 5.5 Specialty chemical and coating sales, 2000
* Total specialty chemical sales estimated at USD 250–300 billion; Source: Lehman Brothers, Chemical Week estimates, McKinsey analysis

First of all, there is still room for considerable consolidation. Most segments – and the industry overall – are still very fragmented. There are over forty large companies with a turnover in excess of USD 1 billion, and several hundred smaller ones. Most of them are competing on a global scale (Fig. 5.5).

In addition, many companies currently have a sub-optimal portfolio mix, including businesses that either have little stand-alone attractiveness (low growth, low margin, or sub-critical size) or are a poor strategic fit, or a combination of the two. In the past, there has been a lack of capital market scrutiny of individual segments within the specialty industry, and strong businesses have been allowed to subsidize weak ones. This has slowed down the restructuring process. Among the portfolios of the largest specialty chemical companies, we found that the truly attractive-seeming businesses often represent only about half the total turnover. The rest are prime candidates for some form of portfolio restructuring.

Finally, the significant valuation differences between companies are likely to make some of them attractive acquisition candidates for the more successful and financially strong players in the industry. Some companies may even, on occasion, appear quite cheap. Also, private equity firms have entered the industry in force and are likely to increase their presence, providing an enabler and catalyst for portfolio restructuring. They have access to plenty of cash (often more than the industry players themselves) and are run by very strong teams, with excellent skills in managing and improving multi-business portfolios (see Chapter 8).

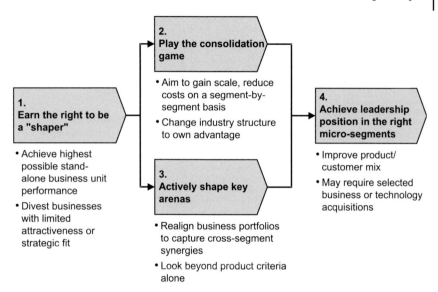

Fig. 5.6 Value-adding portfolio management through arena shaping
Source: McKinsey

Nevertheless, many recent transactions appear to have created little shareholder value because of lack of business overlap (leading to only marginal cost synergies) or because of limited strategic fit in terms of innovation potential or complementary offerings and skills. Typically, this produces almost no top-line effect – the prime driver of company valuation – and may even result in revenue losses in the transition period. The high transaction prices paid in the past few years have done nothing but aggravate the phenomenon.

Against this background, what approach should a specialty company take to optimize its portfolio? Should it pursue absolute scale? Or should it focus management and capital on a narrower portfolio? Should it consolidate and shape industry segments? Should it spin off some businesses? Or should it go for all of the above?

We believe that the future leaders among the specialty companies should make such decisions on the basis of a relentless focus on enhancing the value of their portfolios and harnessing synergies between the individual businesses. Their approach will be based on the following steps (Fig. 5.6):

1. *Earn the right to be a portfolio "shaper"*. Companies should make every one of their businesses as profitable as possible on a stand-alone basis. They will thus gain the maximum value from both the units they retain and those they dispose of quickly because of their limited intrinsic attractiveness and poor strategic fit, and will thereby earn the right to shape industry segments.
2. *Play the consolidation game*. As more and more segments become like commodities, the pattern followed by commodities is likely to be repeated. This means that there is likely to be a continuous process of consolidation in order to gain

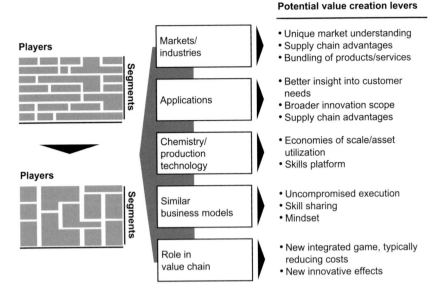

Potential value creation levers

Markets/industries	• Unique market understanding • Supply chain advantages • Bundling of products/services
Applications	• Better insight into customer needs • Broader innovation scope • Supply chain advantages
Chemistry/production technology	• Economies of scale/asset utilization • Skills platform
Similar business models	• Uncompromised execution • Skill sharing • Mindset
Role in value chain	• New integrated game, typically reducing costs • New innovative effects

Fig. 5.7 Five ways to shape arenas
Source: McKinsey

scale, reduce costs, and stabilize the industry structure on a segment-by-segment basis. Specialty chemical companies can improve overall segment performance (and thus their own) by changing the overall industry structure in this way. Companies that play this game will do so in their current segments.

3. *Actively shape key arenas.* Here, the aim is to create value and possibly to expand the market by capturing synergies between different but related chemical segments. Chemical companies can generate value by realigning their business portfolios (by acquisition and/or divestment) along any dimensions that can create real business synergies. It is crucial here to look beyond product criteria alone. For any given group of chemical segments, synergies might be gained from areas like the following (Fig. 5.7):

- The market/customer base – by leveraging, for instance, unique market understanding or supply chain advantages, and offering bundled products or services. Companies that have followed this approach include GE Plastics/Dow in the automotive industry and (the former) SKW Trostberg in construction.
- Applications/solutions – by developing new and more effective solutions or systems (e.g., new formulations) or by exploiting a broader scope of innovation across segments in order to develop new technologies. Examples here are BASF's paint systems (see Chapters 3 and 13), Hercules in water treatment and paper chemicals, or Sivento in surface chemicals.
- Chemistry/production technology – by leveraging common assets and technologies to achieve economies of scale and skills (often in process management) in the production and supply chain, typically to achieve greater pro-

cess efficiency and lower costs. Examples here include Lonza in fine chemical reactions and Rohm and Haas in acrylics.

- Common business model/skills set – by implementing any business model outstandingly well across a range of businesses. This skill-based strategy is one of the most challenging to carry out, but it is nevertheless worth considering for those companies that excel in execution in one or several business models. Examples include Ciba in additives (as a new-product developer) and National Starch and Chemical in food starch (as an applications developer).
- Integration of the value chain – by taking advantage of backward or forward integration in the value chain to reduce production costs or to develop new innovative effects. Nalco in water treatment and Sun Chemicals in inks are two examples of this approach.

4. *Achieve a leadership position in the right micro-segments.* Companies can improve overall performance in those segments in which they really aim to excel through a better product and customer mix, in other words, gaining the largest possible share with the fastest growing, highest margin individual products and customers (see Chapter 13). This might require selected acquisitions of businesses or technologies. Typically, those players that have been able to carve out such leadership positions across micro-segments have achieved much higher top-line growth and margins on a sustainable basis (Fig. 5.8).

In future, we expect most of the specialty segments to experience some important changes, especially those with the potential for consolidation and/or for arena shaping. Examples of such segments include inks, textile chemicals, surfactants,

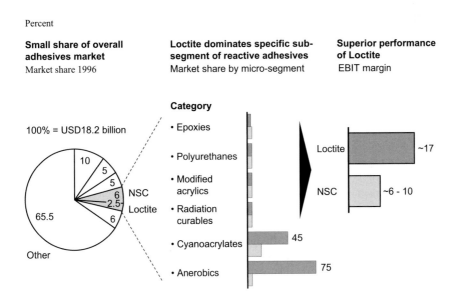

Fig. 5.8 Leadership at micro-segment level critical for superior performance – example
Source: Frost & Sullivan, broker reports, annual reports, McKinsey analysis

adhesives/sealants, paints/coatings, food ingredients, personal care chemicals, and water management chemicals. As with commodities, these changes are likely to be less of a revolution and more of an evolution. They will be driven by the best corporate centers in the industry, by some large and highly skilled financial investors, and (possibly) by some new entrants from the commodity segments.

Specialty chemical companies aiming to be among tomorrow's industry shapers will need to acquire some fundamental new skills, and the process of their development will not always be easy. These companies will, for example, need superior strategic insight into how to create value and how to develop leadership positions at the arena level. Furthermore, they will need financial market-type skills in order to identify, negotiate, and integrate multi-business and multi-company transactions effectively, use their own shares to undertake strategic transactions when market valuations appear high, and perform to the best effect in all aspects of M&A, from deal identification to negotiation, execution, financing, and integration.

Leaders of specialty chemical companies have to be aware, however, that such an industry-shaping journey is a huge challenge and involves high risk. Naturally, such moves demand a large amount of financial investment. But over and above that, major post-merger integration and skill-building programs will be needed to capture the maximum value from arena shaping.

5.3
Achieving Superior Performance at the Business Unit Level

Once a specialty chemical company has shaped its portfolio of businesses, it needs to begin the hard work of maximizing the performance of each of them. In our research, we made an extensive examination of forty industry segments, and identified a shortlist (not necessarily exhaustive) of leading companies and business units which have achieved significant profit growth over a long period of five to ten years and outperformed competitors in doing so (Fig. 5.9). We found that these players executed outstandingly a clearly chosen strategy and business model in line with their industry segments.

5.3.1
Selecting a Single Winning Business Model

Successful players tend to choose one single business model for any given segment rather than attempting to implement a mix (Fig. 5.10). This forces management to focus resources, skills and mindset on the right narrow set of key factors for success which will really make a difference in the market. The winning model for each business segment should be selected on the basis of the nature of the business/chemicals and customer needs in that segment and the stage of maturity it has reached. It then has to be revisited and refreshed regularly as these fac-

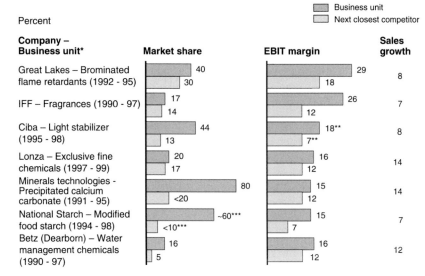

Fig. 5.9 Superior performance at business unit level – examples
* Performance during period of leadership, outside-in estimates; ** Additives; *** US market;
Source: Broker reports, annual reports, McKinsey analysis

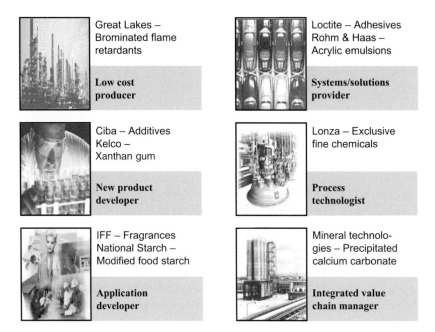

Fig. 5.10 Six distinct business models are driving success in specialties
Source: McKinsey

tors change. Our research has identified six winning business models in the specialty chemicals area:

1) *New product developer:* Such businesses develop (and patent if possible) novel chemicals or formulations that provide additional product performance benefits, or that are lower cost-in-use than the current alternatives. Biotechnology may create new opportunities for this model in certain segments in the future.

2) *Applications developer:* In this model, businesses work hand-in-hand with customers to develop improved or customized products for new applications. Often, these improved or customized products replace another material in the application (e.g., adhesives replacing metal fasteners). Such businesses also provide significant technical services to help introduce and scale up production for these new products.

3) *Systems/solutions provider:* These businesses provide a complete bundle of products and services to customers to reduce customers' total system costs or to increase customer revenues. The services typically help to "integrate" the product bundles to maximize the synergistic effects between products.

4) *Process technologist:* Here, businesses develop (and patent if possible) novel process technologies that allow them and other businesses to produce chemicals at lower cost, at higher quality, or in less time. This technology can be licensed to others for a fee.

5) *Value-chain integrator:* In this model, businesses physically and/or virtually integrate with their customers' value chains. This often involves taking over one or more of the functions that are currently performed by the customer to reduce the customer's total system cost.

6) *Low-cost producer:* Businesses employing this model produce "specialty" chemicals at the lowest cost and provide those chemicals to customers at attractive prices.

As indicated above, the models have to be carefully matched with the nature of the chemicals and needs of the segment's customers and with the segment's state of maturity. For example, a "new product developer" model may have a hard time getting traction in a mature, intensely competitive segment where customers need and expect low cost chemicals. Likewise, a "low cost producer" model may not work in a segment with complex chemical product needs, a high level of interaction between chemicals, significant process cost differences based on different chemical bundles, and competitors that provide a significant amount of technical service. Here, a "systems/solutions" model adds more value.

Assessment and re-examination of the business model is especially important as the segment matures. As a chemical moves from discovery to growth, to maturity, and on to decline, its customer needs, competitive dynamics, and market characteristics change. These inflection points require a refinement (or in some cases total change) of the business model (Fig. 5.11). The key success factor here is the ability to anticipate changing market dynamics in order to be a first mover in developing a new and more successful model.

FUMED SILICA EXAMPLE

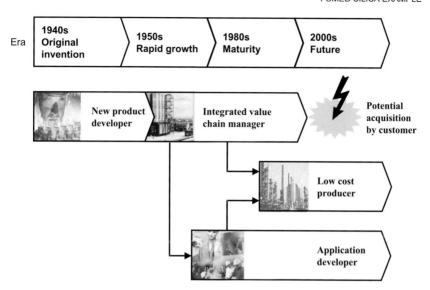

Fig. 5.11 Evolving/expanding business models as market dynamics change
Source: McKinsey

5.3.2
Executing the Selected Business Model Outstandingly

From the analysis of the winners, it appears that the key differentiator between the successful and less successful players in any segment is the quality of their execution of the chosen business model. Improving execution of the optimal business model requires action on three fronts:

Identifying and acting on critical capability areas: these vary significantly between different business models. For example, superior purchasing, ability to run large scale, low cost manufacturing facilities, and an intense focus on sales, general and administration costs (SG & A) are required in the low cost business model. Alternatively, superior and sufficiently funded basic research and development, insightful marketing, and links to external sources of ideas (universities, small R & D houses) are important for a new product developer. Once these capability areas have been identified, successful companies will make major investments in them in order to gain competitive advantage. It will no longer be sufficient to be "good enough" (or even to be "among the best") in areas like technical service, product development, or manufacturing. Rather, in order to be successful, specialty chemical companies will need to secure the talent, build the culture, and obtain the assets (tangible and intangible) required to surpass the competition in the relevant critical areas.

Differentiating management systems and style to match the characteristics of the business model employed by the business unit. Even though it adds complexity, this is essential to ensure that managers are aligned with the key success factors of each model. For example, successful companies will establish cost-focused metrics such as capacity utilization, SG&A as a percentage of sales, and conversion cost per unit as metrics for low cost businesses. These same companies will measure new product development-based businesses by revenue growth and the percentage of revenues generated by products developed in the last five years.

Overhauling skills and motivating talent: this can be done through more aggressive recruiting programs (more sources, higher targets, and more management time spent on recruiting), revamped training programs (to include more field work and greater interactivity between functions), and more attractive compensation packages closely linked to the performance of their individual business units.

In summary, picking the correct business model and executing it better than the competition should enable each business unit to maximize its current performance. That will enable the business to beat the competition in current markets. But what about the future? How will the business unit continue to grow, given mature markets and fierce competition?

5.4
Going for Growth

Growing on a sustainable basis in specialty chemicals is very difficult, as a glance at the industry players' performance across the different segments reveals. Nevertheless, with the traditional sources of demand growth beginning to dry up, it is essential for chemical companies to make particular efforts in this area. We believe that specialty chemical companies will have to expand beyond their current markets and business models by looking outside their current focus on the chemical molecule. Four different options suggest themselves here.

Increasingly focus on service-based business models and develop entirely new service-based businesses. The most exciting business models in today's market tend to be those with a significant service element (e.g., systems/solutions provider, process technologist and value-chain integrator). Companies with these models can differentiate themselves beyond the chemical molecule. In addition, most successful new businesses being created by specialty chemical companies today are service-based. These include operations consulting, asset management, financing, risk management, and safety consulting businesses.

Aggressively seek out and develop new technologies, both within and outside the company. These technologies could include biotechnology product- and process-related innovations, new chiral chemistries and processes, high-throughput screening techniques for rapid new formulation development, and others. Most of the best ideas will come from sources outside the company's laboratories (for example, from universities, R&D houses, entrepreneurs, etc.). Therefore, the best specialty chemical companies will establish important, aggressive technology search-and-licensing functions in the future.

Become a leader in e-business in chosen segments. The Internet is a major discontinuity in the chemicals industry today. Those companies that can successfully take advantage of this discontinuity can expect significantly improved growth from increased market reach, lower costs, and improved customer value propositions. However, barriers to entry in many e-businesses are low, so companies must move fast and establish dominance early to be successful.

Enter new parts of the value chain, typically downstream elements closer to the end consumer. Downstream players often add significant value to the product just before it gets to the end consumer. If specialty chemical companies can enter these parts of the value chain by leveraging their existing business models and core capabilities, they can capture a piece of this additional value. However, this proposition may prove difficult, since the downstream markets are often very different from chemical ones. They are, for example, extremely fragmented, and they have to cope with frequently changing consumer tastes.

5.5
Designing a Value-Adding Corporate Center

In the future, we expect mega-specialty companies to move to a much more decentralized, autonomous business unit structure because this will allow them to optimize performance more actively and to achieve better results in each individual business. This raises a key question: how can the corporate center add value to the sum of its individual businesses to achieve higher shareholder returns in aggregate? The corporate centers of mega-specialties will play an increasingly critical role, especially as industry consolidation is expected to continue and traditional ways of capturing synergies through scale will not work. Only those large specialty companies which are able to implement certain vital elements systematically will be able to achieve a good return to shareholders, add value to the sum of their businesses, and win the industry-shaping game, in particular against financial investors.

Based on our research, we believe that the management of specialty companies should pay particular attention to the following four key elements when establishing a value-adding corporate center.

5.5.1
Putting in Place a Lean, Investor-Type Corporate Center

Leading large specialty chemical companies will be those which put in place a lean, investor-type corporate center and act on levers similar to those used by the leveraged buyout fund managers or private equity players such as KKR, Cinven or Investcorp. We could compare this to a multi-internal LBO approach across businesses. Key levers are:

- Setting demanding growth and profit targets based on a simple set of easy-to-understand performance metrics and financial transparency for each business.
- Ruthlessly focusing managerial and financial investment on the most attractive segments.

- Putting in place the right performance measurement systems and incentive schemes, tailored to the different demands of each market arena/segment. These will often be much more aggressive than even the newest models in "industrial" companies. They will include both real "carrots" for superior performance and "sticks" for poor performance.
- Bringing in leading-edge skills and talent where necessary without compromise, and aggressively weeding out poor performers over time.
- Continuously and systematically reviewing portfolio optimization opportunities and applying best practice M&A skills for successful deal making and integration, avoiding auction fever in M&A and being realistic about synergies.
- Rightsizing the corporate center by keeping only the minimum, highly skilled resources, and where appropriate transferring the rest into the BUs. This has already been done by many companies. One possible benchmark is that corporate center costs should not exceed the 1–2 percent management fees received by fund managers on their managed capital.

5.5.2
Ensuring Differentiated Management of Various Business Units

As described in Section 5.3, business units have to align their organization and skills with the chosen business model and pull all the levers needed for outstanding execution (Figs. 5.12 and 5.13). The corporate center can help to ensure that they remain energized and perform best within the bigger corporation in three particular ways:

1. Use different metrics according to the specifics of the business (e.g., cost position/gross margin for businesses playing a low cost game, share of lead customers for businesses competing through application development). While this seems obvious, it is often not done, as it is much easier and often seems quite natural to apply one single planning process and one performance management system to all businesses.
2. Attract and develop world-class talent and put in place the right teams for each business unit in line with the requirements of the different businesses. Which businesses should your best turnaround managers lead, which ones should be in the hands of your best marketing managers? How well do your BU teams cover the necessary skills required to successfully execute the chosen business model? How well do the individuals within the management teams work together?
3. Apply a different management style and mode of interaction for each business, both to gain better strategic insights throughout the portfolio and to develop the right mindset in business managers (e.g., you would expect to find risk-takers in businesses based on product development, but not in the process technology-based type).

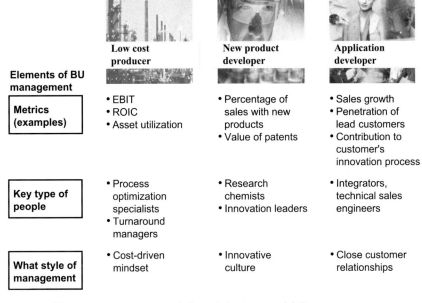

Elements of BU management	Low cost producer	New product developer	Application developer
Metrics (examples)	• EBIT • ROIC • Asset utilization	• Percentage of sales with new products • Value of patents	• Sales growth • Penetration of lead customers • Contribution to customer's innovation process
Key type of people	• Process optimization specialists • Turnaround managers	• Research chemists • Innovation leaders	• Integrators, technical sales engineers
What style of management	• Cost-driven mindset	• Innovative culture	• Close customer relationships

Fig. 5.12 Different management approach for each business model (1)
Source: McKinsey

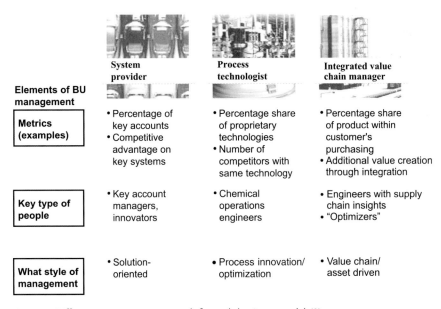

Elements of BU management	System provider	Process technologist	Integrated value chain manager
Metrics (examples)	• Percentage of key accounts • Competitive advantage on key systems	• Percentage share of proprietary technologies • Number of competitors with same technology	• Percentage share of product within customer's purchasing • Additional value creation through integration
Key type of people	• Key account managers, innovators	• Chemical operations engineers	• Engineers with supply chain insights • "Optimizers"
What style of management	• Solution-oriented	• Process innovation/ optimization	• Value chain/ asset driven

Fig. 5.13 Different management approach for each business model (2)
Source: McKinsey

5.5.3
Driving Selected Corporate Themes

Corporate centers are better positioned than business units to initiate and drive selected cross-business themes or initiatives such as procurement, e-commerce or innovation/venturing. They can pool the necessary skills and build competence in areas where BUs do not have critical mass (e.g., business development in China, leveraging biotechnology in chemicals, e-commerce). They can also provide incubators for growth options or build up corporate venture funds, dedicating intellectual and financial resources to selected projects that can help individual businesses secure their long term growth.

5.5.4
Sharing Best Practices Across Businesses

Finally, the corporate center can add value by establishing the right culture and processes for sharing knowledge and best practices between businesses. For example, it can develop a knowledge management system that identifies experts in specific subjects across the company, stores protected and emerging intellectual property, and contains links to external sources of knowledge for the company (e.g., university professors, technical associations, consultants). In addition, the corporate center can sponsor forums for sharing knowledge (conferences, workshops) and encourage individual sharing of expertise through specific incentives (i.e., tie some part of compensation to knowledge sharing). This sharing of knowledge will become more important for specialty chemical companies over the next few years as the source of competitive advantage moves from hard assets to information, knowledge, and special relationships. The corporate center can play an important role in this area, since it does not always come naturally to businesses and individuals within businesses to share their knowledge.

Specialty chemicals are going to face a tougher environment over time. Products and technologies are rapidly maturing and competition between players is getting even more intense. Only those select companies that actually shape their portfolios, regularly refresh their business models and out-execute competitors over time, seek new avenues for growth, and put in place value-adding corporate centers will win.

The extent to which the large specialty companies are able to take control of and shape the market will depend on their ability to outperform other players, be they financial investors or new entrants from the rest of the chemical industry.

6
Chemical Companies and Biotechnology
Rolf Bachmann and Wiebke Schlenzka

Biotechnology is going to transform major segments of the chemical industry during the next decade. The area of agrochemicals has seen biotechnology make major inroads into the traditional business, and further significant changes in the next decade can be expected despite perceived health and environmental risks. In addition, other chemical businesses, where the risks of moving into biotechnology are much smaller, are likely to see biotechnology-based products competing with 30 percent of their trade in the same period.

Chemical companies have a good basis for moving into biotechnology. They have strong manufacturing skills and distribution networks, though they do need to complement these with a new set of skills.

Most chemical companies that are not yet active in biotechnology should delay no longer, as the first movers have already started to reduce competitors' options, and non-movers are unlikely to get a second chance. Some key developments in major areas and some approaches to strategy selection for the traditional chemical companies are outlined below.

6.1
Biotechnology – the Background

Together with information technology, biotechnology has for some time been expected to be the key technology for innovation in the 21st century. However, in many areas it has failed to produce the expected results, and in agrochemicals in particular it has come up against unexpectedly fierce consumer resistance. What, then, are the true prospects in the chemical industry for modern biotechnology, a collection of technologies which analyze and manipulate specific phases in the process of life and which include, among others, genomics, bio-informatics, metabolic engineering and proteomics?

The commercial application of modern biotechnology began in pharmaceuticals at the end of the 1970s when the first biotechnology startup firms were founded. Since then it has been used to design new drugs based on our understanding of the molecular basis of diseases. Hopes for its future range from finding a cure for cancer to gene therapy – the correction of genetic dysfunction by direct modification of a patient's genome.

In agriculture, biotechnology enables the production of more food, and of food of higher quality. Genetically modified crops can give higher yields, require less use of agrochemicals, and have improved nutritional value. In some countries such products are already on the market. In industrial applications, biotechnology permits the development of new, higher-performance materials that (often in contrast to today's synthetic products) are biodegradable.

Despite these great opportunities, however, biotechnology is a risky business for chemical companies. The risks of the technology itself are high because of the long product development times (from five to ten years), the technology's high complexity, and its dynamic nature. Every few years a new technology arises that makes the existing ones redundant.

Moreover, the market risk is high. Consumer acceptance is limited in some areas because of doubts about the long term impact on health and the environment. Even where consumer resistance is less, the opportunities for premium pricing against the existing products are often limited.

Last, but not least, there is the financial risk. The R&D costs are high and can exceed USD 100 million for a single innovative biotech-based product. In addition, in order to ensure that value is captured effectively, expensive downstream acquisitions may be necessary. So why would anyone be willing to invest in this technology?

The answer is that it offers tremendous growth opportunities, and in pharmaceuticals, where its application is more advanced than in any other industry, it has demonstrated that it can deliver high returns. Over the past ten years, most of the truly innovative drugs – those that address an unmet medical need – have come from the application of biotechnology.

Others of the new drugs have been substitutes for existing products. One example is genetically engineered insulin. Heavily discredited by consumer and environmental protection groups in the early years after its introduction in 1984, it has gradually come to substitute for the more traditional animal-based product, in large part because of its higher efficacy and safety.

Many more biotech-based products are currently in the pipeline and expected on the market soon. Most pharmaceutical companies today admit that they are dependent on alliances with biotech startups for the development of new drugs and new ways of treating diseases. The value of collaborations formed between pharmaceutical firms and the biotech industry amounted to no less than USD 4 billion in 1998.

The venture capital industry has been very bullish about biotech, continuously investing approximately 20–25 percent of its total capital in the area over the last ten years. In 1997 this amounted to almost USD 3 billion.

The returns of the publicly-traded biotech venture capital funds – of which we found approximately ten – are comparable to the returns on an average venture capital fund. Fidelity Select Biotechnology, for example, recorded a 26.3 percent average annual return over a five-year period (Fig. 6.1).

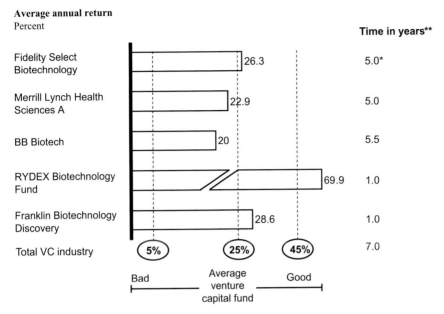

Average annual return
Percent

Time in years**

Fig. 6.1 Venture capital industry succeeds in getting returns from biotechnology
* In 1998, 12-month performance=53.6%; ** Latest year of calculation 1998; Source: McKinsey analysis

The example of the venture capital industry and of a number of biotech startups that have been able to create significant shareholder value (e.g., Amgen, Genentech and Qiagen) demonstrate that the high risks of biotechnology can be managed.

6.2
Attractive in Agrochemicals

Agrochemicals is an attractive biotechnology segment for chemical companies. It seems that the current fears of health and environmental risks are likely to be overcome in the long run, and the potential returns are high. Essentially, the winners here will be companies which already have a very strong base in traditional agrochemicals.

The major segments of this market are still the traditional herbicides, insecticides and fungicides (Fig. 6.2). However, plant biotechnology, though small, is the fastest growing sector of the industry by well over 100 percent, and is likely to have the highest long term impact. In those areas where genetically modified crops with new traits have been introduced, they have rapidly achieved a high market share and have redistributed insecticide and herbicide sales in the process.

By 2010, biotechnology is likely to have significantly transformed the agrochemicals market, with fungicides the only segment likely to remain rela-

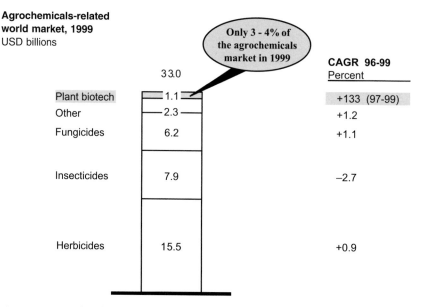

Fig. 6.2 In 1999 plant biotechnology amounted to only three to four percent of the worldwide agrochemicals market
Source: Institut Supérieur de l'Agro-Alimentaire, broker reports, McKinsey analysis

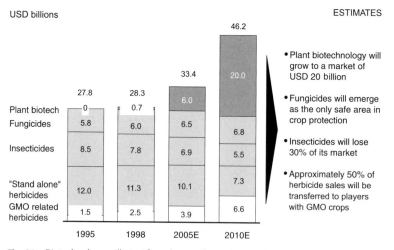

Fig. 6.3 Biotechnology will significantly transform the agrochemicals market by 2010
Source: Institut Supérieur de l'Agro-Alimentaire, broker reports, McKinsey analysis

tively unaffected (Fig. 6.3). A bottom-up market estimate by region and by crop reveals that plant biotechnology in particular is likely to cut the worldwide insecticide market by approximately 30 percent, transfer approximately 50 percent of the herbicide market to players with genetically modified crops, and increase the market segment of genetically modified traits from today's USD 700 million to USD 20 billion over the next ten years.

In 1998, 36 percent of the total soybean planting area in the United States bore genetically modified crops. For corn, the figure was 22 percent, for cotton 20 percent, and for Canadian canola 60 percent. The figures increased to more than 50 percent for soybean, 30 percent for corn, and more than 30 percent for cotton in 1999. Farmers using genetically modified seeds achieved approximately 10 percent higher net returns due to their higher yield and the lower use of agrochemicals. Insect-resistant, genetically modified cotton reduces the use of insecticides on average by 50 percent.

In the herbicide market, more than USD 250 million of the sales of herbicides in the USA were transferred in 1998 from "stand alone" herbicides to herbicides that could be combined with a genetically modified crop. This redistribution of herbicides puts traditional agrochemical businesses at risk. Companies where herbicides account for more than 50 percent of the total revenues and that have a high market share in the USA are already suffering. The biggest short term losers are players that used to have strong sales in those areas (like soybean and corn) where herbicide-resistant crops have been particularly successful.

The early success of plant biotechnology has recently been threatened by increasing consumer resistance. While the introduction of genetically modified crops has been very successful in the USA, Canada, Argentina, Brazil, Australia and China, consumers in Western Europe and Japan have heavily opposed prod-

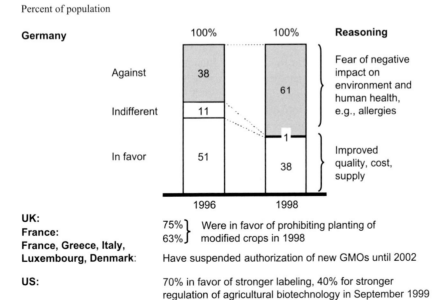

Fig. 6.4 Biotech food and plant products increasingly lack consumer acceptance
Source: The Economist, June 1999, Europa Chemie 12/99, EMNID, McKinsey analysis

ucts from genetically modified crops and have effectively blocked their commercialization. In the UK in 1998, for example, 75 percent of the population was in favor of prohibiting the planting of modified crops. Moreover, the situation seems to be getting worse rather than better. In Germany, the percentage of the population against the planting of modified crops rose from 38 percent in 1996 to 61 percent in 1998 (Fig. 6.4).

The fear in Europe of negative long term effects on human health and the environment has now started to move to the USA. A recent US consumer survey conducted by StrategyOne in September 1999 demonstrated that consumers want more labeling and stricter regulation of food containing genetically modified organisms. Almost 70 percent of the survey respondents said the US government should require companies to provide more extensive labeling of ingredients in GM food and 40 percent said agricultural biotechnology should be regulated more closely. Furthermore, the survey showed that consumer knowledge of gene-modified food is still limited: 62 percent of 1017 randomly polled adults were unaware that about half of the nation's food contains gene-altered ingredients.

In addition, 30 percent of the total US crop sales are exported to Europe, so that American farmers are beginning to fear that they will not be able to sell their GM crops and are reluctant to plant them despite their agro-economic advantages. The lagging consumer acceptance is sure to cause a slowdown in the penetration of genetically modified crops over the next few years.

However, all the scientific data gathered over the ten years since the first field tests on the genetically modified organisms has not revealed any higher risks for consumers and the environment than exist with today's agriculture. Also, biotechnology provides the scientific basis for closing the "food gap" that arises from the decline in arable land on the one hand and population growth on the other. For these two reasons, we expect the negative feelings towards genetically modified crops to decrease significantly after 2005, when the first modified crops will have been on the market for ten years and it will be evident that harmful effects are controllable.

What are the strategic options for agrochemical players in this changing environment? The key parameters here are traditional crop protection and plant biotechnology, where strategies have to be based on the companies' ability to exploit the different types of improved plant traits. These new traits improve the agronomic crop characteristics, for example, pest/disease resistance, reduced fertilizer demand, better usage of soil minerals, less demand for water (input traits) or the characteristics of the agricultural products, for example, higher vitamin content, no allergenic substances (output traits).

The winners will have to be strong in both traditional crop protection and plant biotechnology, and will be able to combine both skills effectively. They will have operational excellence in traditional agrochemicals and will use the returns from the traditional business to invest in plant biotechnology as a medium term growth platform.

They will capture value from new genetically modified crops with improved traits by forward integration into the seed business. In addition, understanding plant genomics will help them to design innovative new agrochemicals as well as to lock in herbicide sales.

Without strengths in plant biotechnology, pure traditional agrochemicals players will have to refocus on niche segments, but will have only limited opportunities for growth.

Leaders in plant biotechnology only, however, with a weak position in crop protection, should focus only on the output characteristics, for such companies are not likely to be able to take advantage of the shift in agrochemical sales. As the traditional business is weak, such companies face high risks from delayed consumer acceptance.

Compared to other chemical businesses, though, agrochemicals are highly attractive. Key agrochemical companies are making a drive for operational excellence in traditional agrochemicals (through consolidation) and are starting to invest in biotechnology, and they can be expected to achieve a return on sales of 20 percent. We are therefore currently seeing a new top league of agrochemical players arising.

6.3
Other Chemical Businesses: Fast Moves Needed

The application of biotechnology in other sorts of chemical production is not new. Some basic and specialty chemicals (for example, ethanol and starch derivatives) have always been produced by classic fermentation processes. These, however, are traditional niche applications. Modern biotechnology has the potential to be fundamentally and broadly innovative in chemical processes and products, right across the board. Its application is likely to grow more than tenfold over the next decade. These areas are less risky to enter than agrochemicals, but for a variety of reasons they are also moving faster, so it is important for chemical companies to choose and implement their course fast. It appears to be more of a risk for chemical companies to fail to invest in biotechnology, because they may then be left standing and will miss one of the biggest growth opportunities of the 21st century.

Genetically designed enzymes, cells and organisms will produce or modify chemicals in a way that even today is unimaginable. We cannot be sure about how things will be in ten years' time, but one thing is certain: imagination will be a key driver of the future. Companies with a vision and the aspiration to achieve it will then drive the speed of development. For example, the chemical industry only developed environmentally advanced processes when it inspired this move itself, after major incidents of pollution.

Examples of biotech-based processes and products that are already to be found are biocatalysis and biomolecules in fine chemicals, biopolymers as substitutes for synthetic polymers, enzymes and modified additives in specialties, and modern fermentation as a production process for basic and intermediate organics. The market for biotech-based products (excluding traditional fermentation in, for example, ethanol production) accounts today for only two percent of the total chemical market, which is worth approximately USD 25 billion.

In those product segments in which biotechnology plays a role, however, it has been able to reach a dominant position within a very few years. It has, for example, gained a 100 percent market share in some organic acids, that is, citric and L-lactic acid, vitamins such as vitamin B2, amino acids, that is, glutamic acid and phenylalkaline, and others like nucleotides, enzymes or xanthan gums. The main reason for this high level of substitution is the significant cost advantage of biotech-based production processes. In riboflavin, for example, the biotech-based process almost totally replaced the market for products made by chemical synthesis within four years. Production costs were more than 50 percent lower, and investment in new capacity required 40 percent less capital.

We estimate that biotechnology will be competing with approximately 30 percent of the total chemical market by 2010 on the basis of lower cost and/or superior product features. Approximately 10 to 20 percent of basic and intermediate chemicals could be affected by production through modern fermentation. Specialties will be replaced by enzymes and natural flavors, pigments and additives. Polymers will face competition from biopolymers that are competitive in price with both polyester and nylon.

This bottom-up estimate correlates with the sales expectations of some leading players: DuPont, for example, announced in September 1999 that they expected 25 percent of their material sales to come from biotech-based products in 10 years' time.

The advantages that made biotechnology a winner in the areas it has already entered apply to many other chemical businesses:

- It is cost-effective. Biotech-based processes have substantially lower capital and manufacturing costs.
- It allows for greater flexibility because the minimum plant size to achieve economies of scale is much lower.
- It is more eco-friendly because there is less waste and less energy consumption, and its products are often biodegradable.
- It is more sustainable because it relies more on renewable resources.
- It is potentially revolutionary because it holds out the promise of features that are unknown in existing synthetic materials.

Three examples demonstrate these advantages in more detail.

1. Vitamin C

   The semi-synthetic production of vitamin C is rapidly moving to a full biotech process. Vitamin C (ascorbic acid) is an important segment in the worldwide vitamin market with a market share of approximately 20 percent. Its worldwide sales amounted to around USD 0.5 billion in 1999. The traditional route to vitamin C is a multistep process involving chemical and fermentative steps. It starts with the catalytic hydrogenation of D-glucose to D-sorbitol, followed by the fermentative oxidation of D-sorbitol to L-sorbose, which is then converted by a multistep synthesis to 2-keto-L-gulonic acid, the key vitamin C intermediate, and finally to ascorbic acid. This semi-synthetic process was developed by Reichenstein and Grussner in the thirties and is still today – with only slight modification – the most common process in vitamin C production. The high complexity of the production process and the importance of the market triggered ongoing research into a fully biotech-based process. Applying gene technology, three well-known biotech startups have recently been successful in establishing such a route. Genentech created through the application of biotechnology a gluconate reductase enzyme that is capable of producing the vitamin intermediate directly at high yield and eliminates four synthetic steps at a stroke. At the same time, Cerestar developed a direct route for sorbitol to the vitamin C derivative and started the first commercial plant running in 1999 in collaboration with BASF and Merck. Finally, in 1999 Genencor announced the development of a wholly biotech process to vitamin C that eliminates several steps from the traditional one and is totally aqueous. The process achieves significant cost savings through lower capital costs – smaller and more efficient factories – and higher yield and productivity. Furthermore, the impact of this new process is much broader, as it will be applicable to more fossil carbon-based chemicals and will allow their synthesis from renewable carbon from agriculture at lower cost.

2. Biopolymers

Polylactic acid and 3 GT (3 carbon glycol terephthalate) are two biopolymers that are already being successfully produced in pilot plants. Their manufacturers (Dow for PLA and DuPont/Genencor for 3GT) expect to reduce the production cost to USD 1 per kilo or even lower by 2003. These biopolymers will then become competitive with polyester and nylon chips. Dow and DuPont have both begun the construction of large-scale biopolymer production plants.

In addition, biopolymers (in contrast to most synthetic polymers) are biodegradable. A PLA yogurt cup that is already on the market disappears within 40 days.

3. Biosteel

The third example is biosteel, a revolutionary new material that can open up large, completely new markets. Anyone who has ever touched a spider's web knows that spider silk is a very strong, elastic material. Now scientists are able to isolate and clone spider silk genes. In 1999, they were able to transfer these genes to goats which secrete the proteins in their milk. The next research step is to develop a technical spinning process. Spun "biosteel" could then be used, for example, as a substitute for metal in the construction of earthquake-resistant bridges, or as a material for artificial blood vessels or implants.

Current R&D in a number of other areas indicates that biotechnology is about to have a major impact on many more businesses. In areas as far apart as water management and adhesives, biotechnology applications are either in the laboratory or in pilot plants (Fig. 6.5).

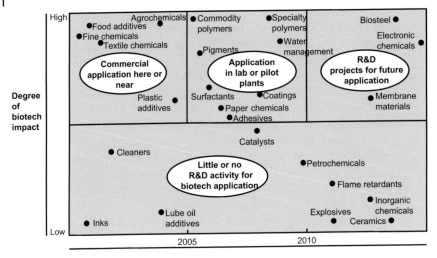

Fig. 6.5 Status of biotechnology in 1999 – preparing to attack a number of additional chemical businesses
Source: McKinsey analysis

But biotechnology is not superior to chemistry in all areas, and it is not about to make chemistry as meaningless as alchemy. The idea must be to combine the best of both worlds. It has already been shown that one biochemical step in an otherwise pure chemical production process can bring about fundamental economic improvements.

The risks of moving into biotechnology are much smaller in other chemical businesses than they are in the agrochemicals sector. Consumer acceptance is often high because of the environmental benefits, and the R&D costs are lower because of the shorter replication cycles in microorganisms and fewer regulatory hurdles.

Moreover, chemical companies have a good basis for moving into biotechnology. They have strong manufacturing skills and distribution networks, though they do need to complement these with specific biotechnology skills, alliance management skills, and effective financing and risk management skills.

Therefore, most chemical companies that are not yet active in biotechnology should act today, as the first movers have already started to reduce competitors' options. Areas where firms need to move quickly and aggressively are textile chemicals, fine chemicals, and additives. In contrast with agrochemicals, with its slow consumer acceptance, non-movers are not likely to get a second chance.

6.4

Defining the Directions for Moving into Biotechnology

To make a successful move into biotechnology, chemical companies with a mixed portfolio of businesses, be they large conglomerates or narrower life science companies, need to work out which of their many individual businesses are suitable candidates, determine which ones to select for a broader portfolio, and trace out a route for making the transformation.

6.4.1

Performing a Diagnostic for Individual Chemical Businesses

To find out where to move, companies should analyze their current businesses and existing capabilities and consider the threats and opportunities of biotechnology along three dimensions: a) the level of the impact of biotechnology on the business, b) the timeframe for the expected technology shift, and c) the importance of the business to the company. Based on this analysis, they should develop both a corporate strategy and a strategy specific to each business unit, addressing the threats and the options for growth in biotechnology. If a core business is expected to be seriously threatened within the near future, for example, as was the case with herbicides in agrochemicals, the company has to move fast or lose significant market share.

6.4.2

Outlining the Strategic Options for Biotechnology Portfolios

In addition to individual businesses, chemical companies have to consider their biotechnology strategies from the point of view of a portfolio and the synergies it offers. The example of the life science companies shows that biotechnology does not contain synergies right across the board, but there are important ones which conglomerates or life science companies can use as a basis for portfolio selection.

A general move away from classical chemical businesses started at the beginning of the nineties, when chemical conglomerates spun off their chemical businesses and announced a shift to more profitable, non-cyclical "life science" businesses. The logic of this move was to restructure the portfolio into businesses with synergies in the market as well as in technology. The technology synergies were mainly expected to come from the cross-business application of biotechnology, essentially in pharmaceuticals, nutrition and the agro-businesses.

Analysts were excited about the shift, and the early life science companies outperformed the chemical conglomerates and put in a performance similar to that of the top pharmaceutical companies. As a result, many other chemical conglomerates followed the trend with their own definitions of a life science portfolio.

Even businesses like fine chemicals, anti-infectives, food specialties, bakery ingredients and specialty intermediates were bundled under this heading. However, most of the life science companies did not fulfill investors' expectations. The expected technology and operational synergies turned out to be limited, and the individual business units were too often run as fully independent businesses. As a result, the total returns to shareholders slowed down to the level of the chemical conglomerates. We therefore expect that other so-called "life science" companies will follow the example of Novartis and AstraZeneca (newly formed Syngenta) and split again into pure pharmaceutical and agrochemical companies.

From today's point of view, there appear to be three areas in which multibusiness companies should seek the potential for synergies: technology, market and value chain focus – but not necessarily all three together (Fig. 6.6).

Technology focus. There are three very different options here, determined by the natural diversity of living organisms: human, plant and micro biotechnology (i.e., the use of microorganisms and biomolecules). As mentioned above, however, the synergies have to be sought within each of the three areas and not across the three, since their application and the individual technologies are in general very different.

Market focus. Pharmaceuticals, feed/food, consumer and industrial products will remain very different markets for the next ten years. Beyond this period, however, they might well converge; for example, pharmaceuticals and food might grow together as a result of the development of health foods.

Value chain focus. Likely candidates here are R&D, manufacturing and marketing. Traditionally, most companies cover all these positions in their core businesses. For example, most pharmaceutical companies do their own research, manufacturing and selling of drugs. However, more and more companies are starting to focus on individual positions in the chain: biotech startups, for example, focus on R&D, while toll manufacturing companies have effective and efficient manufacturing as their core competence.

A comparison of a life science pharmaceutical business with a life science agrochemical one along these dimensions makes it clear that the synergies are constrained by, for instance, the different technology and market focus.

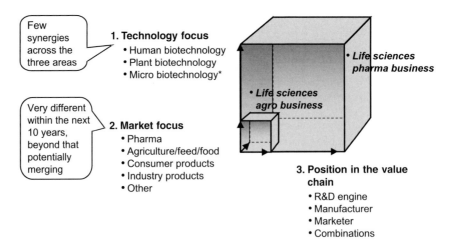

Fig. 6.6 Strategic options in biotechnology
* I.e., focus on microorganisms and biomolecules; Source: McKinsey analysis

The optimal biotechnology portfolio consists of businesses that show synergies along at least one dimension. For today's life science companies and chemical conglomerates, we see five major strategic options which combine biotechnology and market focus. Although conglomerates do have the option in theory to occupy biotechnology space in several of these, it seems that their best chance of success lies in concentrating on one of them:

1. Human/animal biotechnology combined with the pharmaceutical/animal health market.
2. Plant biotechnology and the agriculture/feed/food market, which implies occupying a strong position in genetically modified crops (seed) for feed and food and in agrochemicals.
3. Micro-biotechnology and agriculture/feed/food, with a focus on fermentation and feed/food additives and ingredients.
4. Plant biotechnology and consumer/industrial products. Companies choosing this option will concentrate on genetically modified, plant-derived additives and raw materials, for example, modified natural fibers.
5. Micro-biotechnology and consumer/industrial products, that is, the market for fermentation-based raw materials and additives.

6.4.3
Making the Transformation

To manage the transformation to biotechnology, the chemical companies need to open up to the external community to identify opportunities and acquire biotech knowledge and expertise, since most of the relevant technologies are developed by public research institutions and startups (see Chapter 9).

In areas where competitors have already made important moves, acquisition of small first movers might be the only entry option for the latecomers. In managing internal biotech R&D, companies should use corporate venturing. This is an effective tool for gaining access to the new technologies and also for stimulating internal innovation (see Chapter 9). As the performance of the biotech venture capital funds demonstrates, the venture capitalist approach can allow the risks in biotechnology to be managed successfully.

In the agrochemicals sector, where the application of biotechnology in chemical businesses is most advanced, the leading players have already demonstrated successful transformations along these lines that have led to a string of new ideas and new businesses: key agrochemical players, for example, have set up corporate venture capital funds and created strong cooperations with startups.

Making a commitment to biotechnology will profoundly affect the organization and mindset of the typical chemical company. Nonetheless, chemical companies, which already have strong operational skills and customer relationships, are well positioned to manage the risks effectively and achieve profitable growth through biotechnology.

7

The Impact of E-Commerce on the Chemical Industry

Ralph Marquardt, Jan-Philipp Pfander, Boris Gorella, and David McVeigh

As in many other industries, Internet startup companies staked out positions in chemicals as early as 1997, and initially it looked as though they would threaten the place of traditional chemical companies. It now seems, however, that the incumbents cannot be dispossessed that easily: their strong supplier-customer relationships and an unprecedentedly fast response to the new development places them in a very strong position. Nevertheless, though the players may not change markedly, e-commerce will transform the buying process and those functions which are directly at the customer interface: sales, marketing and supply chain management. Sales and delivery will be much more efficient and much more customized, and it will become feasible to serve smaller customers more cost efficiently. How chemical companies take advantage of the opportunities will depend on the industry segments they play in. They are likely to focus on managing the customer interface for a selected customer portfolio, but will bet on different channels initially to allow customers to "choose" the winner. Their main challenges will be to develop a set of viable sales channels, secure their fair share of the value generated, and manage the fast transition smoothly. Now that the hype of the first few months has faded, e-commerce activities in the chemical industry are already showing signs of a fundamental change in interaction between the established players, which will alter the competitive positioning in the industry over the next few years.

7.1
Winning Control in E-Commerce

E-commerce was originally introduced to the chemical industry by fast-moving startup companies that impressively illustrated the potential of e-commerce for this fragmented industry and were well funded by venture capitalists and enthusiastically endorsed by the capital markets. Many of these attackers started as infomediaries, websites that provided content and community functionalities, which then developed into electronic marketplaces offering some of the chemical companies' traditional transactional capabilities, e.g., order entry, order tracking, a sophisticated catalog and product comparison capabilities, and invoicing. SciQuest, for example, is a website offering life science research supplies with the value proposition of being a one-stop shop. SciQuest provides researchers with new and convenient ways of purchasing their laboratory supplies directly over the web, and facilitates the payment process. The PlasticsNet website targets the injection molding industry, offering plastic resins from different suppliers as well as machinery. All of the new players are struggling to attract enough liquidity to survive among the 100 plus chemical e-marketplaces currently operating or expected to go live by the end of the year 2000.

Incumbents reacted to the emergence of electronic marketplaces by building their own websites, many of which initially only provided basic product information. Chemical companies like Eastman and Dow quickly began building proprietary extranets offering detailed product information and transactional functionalities. However, the incumbents' extranets faced the challenge that the customer still had to use a large number of individual supplier websites to purchase his raw material portfolio and to gain market transparency, which forced them to focus on larger accounts or to provide services far superior to industry standards.

In a third wave of activities, incumbents have set up industry consortia addressing the incomplete value proposition of multiple supplier websites and defending the ownership of the customer interface against attackers. One example is Omnexus, an industry consortium originally set up by BASF, Bayer, Celanese (Ticona), Dow and DuPont, which is targeting the plastic injection-molding industry. Another example is Elemica, an industry consortium which was founded initially by eight leading chemical companies and distributors, and which covers a broad range of chemical industry segments. One prominent advantage of these industry consortia is their ability to leverage the strong brand name and customer contacts of their founding members to gain liquidity fast. However, they are faced with several problems. For a start, they have to convince customers that they are actually neutral; they also have to align the support of the founders behind the new independent entities in a meaningful way and convince capital markets that their value proposition can become profitable over time and will grow beyond the initial segment focus.

It is likely that many of the attacker e-marketplaces which have not yet gained much liquidity will be consolidated by large incumbent-originated consortia such as Elemica and Omnexus or by the extranets of major incumbent players, including large distributors. However, some specialized players will carve out their niches in some industry segments or morph successfully into the second best option of being only an e-commerce service provider. Notably, there have been few announcements from new attackers since the foundation of these consortia.

As a logical extension of these extensive incumbent activities, it seems likely that most customer/supplier interactions will be e-enabled in some ways within a few years. Leading financial analysts like Goldman Sachs and Banc of America predict that 20–30 percent of all business-to-business (B2B) chemical sales in 2005 will be conducted online over the Internet. Many industry quotations from leading chemical companies point in the same direction; GE Polymerland predicts that 40 percent of their sales will be effected online in 2000, DSM 50 percent in 2003, BASF forecasts 40 percent in 2002 and Dow 80 percent in a few years. Given this rapid development, e-commerce is at the top of every CEO's agenda, and they are asking themselves some urgent questions:

- What will the value creation potential and risks for my industry segments be?
- Which strategic moves and position do we have to take to maximize value creation from e-commerce investments?
- How can we overcome the transitional challenges faced by an incumbent organization while its business becomes e-enabled?

7.2
The Creation and Redistribution of Value

In general, e-commerce will affect every aspect of a company's interaction with the outside world of suppliers, customers and competitors. The effects crystallize out into three main forces for value creation and distribution: cost reductions due to increased process efficiency, market share gains or losses due to new value-added services, and margin erosion or gains due to increased price transparency. In view of the extraordinary volume of activity, the main source of differentiation will lie in excellent marketing and sales and the benefit in value redistribution due to transparency, whereas cost reduction will not be a differentiating factor because every player in the industry will be able to do it. The relative importance of these value creation forces depends heavily on which industry segment a company plays in. Although the net effect for the chemical industry might only be slightly positive, since the total cost reduction will probably equal the margin erosion, the profitability gap between leaders and laggards could potentially widen by approximately 5 percent of ROS (Fig. 7.1).

Cost reduction: E-commerce can completely automate repetitive day-to-day processes like order entry and order tracking as well as routine information retrieval, resulting in potential cost savings in sales, general and administration (SG&A) cost sometimes exceeding 20 percent. The cost of internal customer service representatives, in particular, can be substantially reduced by the automation of order management. On the purchasing side, e-commerce will also help reduce the complexity of purchasing indirect goods such as maintenance, repair and operations (MRO) material. The increase in supply chain information, for example, through collaborative planning or vendor-managed inventory, will lead to additional cost reductions due to reduced inventory levels; it is widely estimated that inventory costs can be reduced by 10 to 20 percent.

Shifts in market share: Generally speaking, e-commerce will revolutionize the marketing and sales process by improving the quality of information about customers and products. An e-commerce solution will revolutionize the ease with which external information can be retrieved and internal information generated, giving companies the opportunity to cost-efficiently customize the interface and the service offering to narrowly defined customer segments or even single customers. This will make it possible to sample information about the buying processes, in-

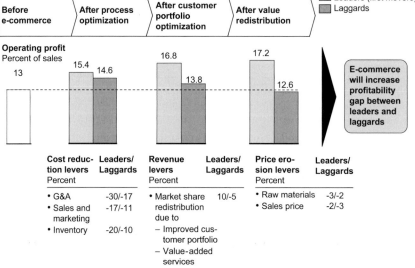

Fig. 7.1 Potential profit improvement levers related to e-commerce
Source: Deutsche Banc Alex. Brown estimates, McKinsey analysis

formation needs, and access levels of individuals as well as their transaction history, and to use this information to customize every point of contact – not only websites, but also sales force and call center activities. In the past, for example, providing customers with regular technical information updates has been a major challenge. In the specialty chemicals industry, products are widely used in formulations which offer a broad range of different applications. By offering access to their formulation databases, suppliers can offer their customers a high value added since this can reduce their internal formulation cost. The wide variety of web-enabled value-added services will allow some suppliers to be truly distinctive and thereby gain market share. These new value-added service options could be extended to the point where virtually new service-driven businesses are created. Company interviews indicate that revenue gains of up to 10 percent can be achieved in markets with a fragmented customer base by offering superior value-added e-services.

Changes in margins: Overall, e-commerce will lead to an increase in price and market transparency due to a higher frequency of requests for quotes (RFQs) because of the improved convenience of the RFQ process and through value-added functions such as availability to promise (ATP). In addition, catalog engines will offer special tools for better product comparison, eventually driving further standardization. Since current pricing in most industry segments reflects the past negotiation skills of the customer rather than the actual total value added, increasing transparency would result in potential erosion of margins. This effect may be even further amplified by the changing competitive dynamics resulting from reverse auctions. However, companies may benefit from auctioning their off-spec material and may achieve higher prices due to the increased market transparency. In addition, marketplaces may use e-commerce to improve their communication of price differences related to product and service quality, by creating increasingly differentiated quality indices that can be used in formula price-based contracts. Chemical companies will benefit from this price erosion in raw material procurement, but will be threatened on the selling side. The net effect on a company will depend on its position in the value chain. A chemical producer which is heavily dependent on naphtha or olefines as raw materials will be unlikely to gain from additional price erosion on the purchasing side because of relative market transparency and the high consolidation of the supplier industry. If, however, that chemical producer is facing substantial competition in his segment, the price erosion potential on the selling side will exceed that on the purchasing side, leading to a substantial reduction in profit. Thus, the effect of price erosion will vary substantially from segment to segment, with margins in some segments getting pinched. Additional consolidation may take place in the segments that suffer most, leading to a rebalancing of bargaining power in the value chain.

The type of supplier-customer relationship and the degree of supply chain complexity are likely to determine the opportunities and the recommended strategic approach to value creation. To map the opportunity and the recommended strategic approach to value creation, it is useful to classify segments by the type of sup-

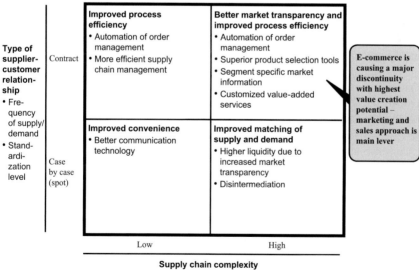

Fig. 7.2 Basic segments within the chemical industry – degree of e-commerce impact
Source: McKinsey

plier-customer relationship and the degree of supply chain complexity (Fig. 7.2). Basically, two types of supplier-customer relationship can be distinguished: the contract relationship is characterized by high switching barriers; usually the product is not highly standardized and/or a high volume is needed relative to market liquidity, so that frequent purchases are made from one supplier. The case-by-case relationship, in contrast, can be defined by low switching barriers, high product standardization and an irregular purchasing frequency. The supply chain complexity is directly related to the number of suppliers, customers and distributors as well as the number of production sites. For each of the resulting four industry segments e-commerce has a different value proposition and value creation potential:

Contract relationship/low supply chain complexity: E-commerce creates value for buyers here by improving the efficiency of their procurement department, for example, through automation of the order management process by a direct ERP-ERP (enterprise resource planning) connection. Suppliers also benefit from deep supply chain integration, because of process efficiency gains in order management (i.e., the customer service function), and because they can offer value-added services such as VMI (vendor-managed inventory). In particular, they can benefit from the improved supply chain transparency and reduce their inventory levels substantially. The overall impact of e-commerce in this segment will be moderate, however, because the cost savings potential is limited. An example of a chemical company which uses supply chain integration extensively is BOC, an industrial and specialty gases producer, which claims to have achieved substantial cost reductions by introducing VMI to both customers and suppliers at more than 100 of its branches.

Contract relationship/high supply chain complexity: In these segments, e-commerce helps purchasers to make substantial reductions in the complexity of their procurement processes by providing intelligent product selection tools and order management support. In addition, customized value-added services like formulation recommendation engines support the purchaser's product development. The supplier, on the other hand, not only benefits from efficiency gains in his order management process but also obtains a cost-efficient channel for customized value added services; for example, e-technical service based on an internal knowledge management system. A further advantage on this side is the increased transparency of customer data, which supports the marketing department in adapting the service offering to specific customer segments. These customized value-added services will allow leading suppliers to gain both competitive differentiation and market share. E-commerce will create a true discontinuity in this segment by fundamentally changing the marketing and sales process. Therefore, the gap between leaders and laggards will widen dramatically. Consequently, a high level of e-commerce activity can be observed in these segments, and a wide variety of business models are currently emerging. GE Polymerland is a prominent example of a chemical company which offers some of these new services, for example, a custom color-matching service dedicated to injection molders, an extensive library of online technical information, or e-seminars covering product application-related topics. The industry consortium Omnexus, which targets the same segment, is another example here.

Case by case relationship/high supply chain complexity: E-commerce creates value on the purchasing side here by increasing market transparency and thus potentially reducing prices, while suppliers gain increased access to untapped customers. As mentioned above, suppliers could even potentially achieve higher prices for their off-spec material due to this increased market transparency; steel makers, for example, have achieved up to 13 percent higher prices for off-spec steel. Overall, e-commerce will create moderate benefits in this segment. An example of a player in this market is ChemConnect, a website offering products ranging from fine chemicals to plastics on the basis of a bid and post system.

Case by case relationship/low supply chain complexity: E-commerce creates value for suppliers and purchasers by providing a more convenient trading channel for large volumes of commodity chemicals like benzene, methanol, and so on. Potentially, new financial products for risk management may also be created. Because of the high consolidation on both the supplier and the purchaser side, no substantial price erosion is likely. Overall, e-commerce will have a very low impact in this segment, since existing business processes will not be substantially changed by the new tools. A typical player in this segment is CheMatch, a website specifically targeting the basic chemicals industry.

In the following section, we will discuss how best to leverage e-commerce strategically and operationally to maximize the value at the customer interface, a focus which is also relevant to its mirror image of purchasing. Process cost reductions will not be separately addressed because of their lack of differentiating power.

7.3
Strategic Questions for Top Management

The high uncertainty about winning models in the industry today has induced chemical companies like Bayer, DuPont, or Dow to bet on several partially competing e-commerce solutions, offering a host of functionalities, with different partners at the same time to secure their ownership of the customer interface. However, in order to develop a robust e-commerce strategy a company's top managers have to assess what value propositions will be the most successful in the long run, as well as which e-functionalities/solutions and what integration into the existing business will be needed. In addition, they have to find the right partnerships for success and decide on the ownership options required to secure the highest share of the value created.

We believe that in the long run it will not be necessary to discuss buyer- and seller-centric models separately: these two models seem likely to converge over time because the winning models will reflect the best value creation for sellers and buyers alike. This is why seller-originated consortia such as Omnexus very much stress their independence and will use their freedom to create the maximum customer value, even though this might not be in the short term interest of the founders, to ensure success against competing attackers. In addition, chemical companies will often buy and sell through the same marketplace, such as Elemica, and therefore require a consistently unbiased e-commerce solution. However, in some cases marketplaces may originate from a seller-centric or buyer-centric position, depending on the industry structure and the liquidity-building strategy.

To determine the best options, a company's portfolio of customers/industries served should be mapped according to the industry matrix shown in Fig. 7.3. This exercise makes it possible to map the current situation as well as to anticipate the future changes in the supplier-customer relationship which may be caused by e-commerce. For example, customers may shift from contract to case by case purchases as markets become increasingly transparent and products more standardized. Supply chain complexity will determine how much value a customer can expect from increased transparency, which will encourage him to seek solutions where a wide range of offerings can be compared. Given the growing number of choices available, customer behavior is likely to polarize. E-commerce will either intensify cooperation between companies, because key buying factors such as reliability of supply and stability of product quality will lead them to improve the mission critical relationships with their suppliers, or will foster standardization to reduce switching barriers. Companies therefore have to be very honest with themselves and assess the likely behavior of their customers carefully in order to be able to predict the future winners. However, for many segments the contract relationship will dominate the industry/customer portfolio, as indicated by the generic example of a polyolefine application portfolio.

For each segment in the industry matrix, a preferred e-commerce solution (an e-channel) will evolve which exhibits a distinctive set of e-commerce functionalities (Fig. 7.4). Four basic e-channels can be defined: *extranets*, which serve as a

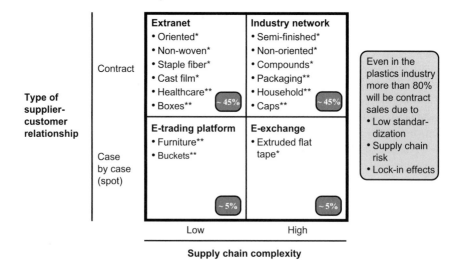

Fig. 7.3 Example of channel choice in plastics industry based on industry segment analysis
* Extrusion; ** Injection molding; Source: McKinsey

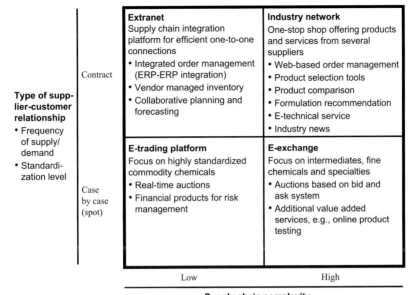

Fig. 7.4 E-channel-specific e-commerce functionalities/solutions
Source: McKinsey

supply chain integration platform for efficient one-to-one connections, *industry networks*, providing a one-stop shop for products and services from several suppliers as well as transparency, *e-exchanges,* focusing on creating transparency in fragmented markets for well-defined chemicals, such as some intermediates, and *e-trading platforms*, focusing on trading in highly standardized commodity chemicals with low supply chain complexity.

Using this basic strategic framework, companies need to develop their value proposition together with the right set of partners to secure the maximum share of the value created.

Extranets. The natural owner of these is the incumbent producer. The focus should go beyond the obvious, such as transaction capabilities, to create solutions which offer additional value to the customers. A new intermediary cannot add any value in this segment and is likely to fail because of the existing strong supplier-customer relationships. Alternatively, an industry consortium owned by the incumbent producers may evolve which provides IT standardization and superior supply chain integration capabilities. Overall, the role of the industry consortia in this segment will be to standardize the IT interfaces between chemical companies. We believe that the IT architecture of the individual extranets will converge over time, thereby weakening the initial value proposition of the industry consortia. Elemica, for example, has announced that it will offer ERP-ERP integration capabilities in its second release. If an incumbent producer has significant sales in this segment, it will have to strive for ownership by building its own proprietary extranet as a no-regret move. In addition, it may use or invest in an industry consortium as an alternative e-channel.

Industry networks. The natural owners are the incumbents in this case, too, because they own the existing customer interface and can provide the e-channel with the necessary liquidity to make its value proposition work. The industry networks may create the highest value from access to the customer interface, which will eventually be developed into a close relationship. Therefore, the emergence of this e-channel is a true discontinuity for the marketing and sales approach which will fundamentally change the way business is conducted in the relevant segments of the chemical industry.

An industry network can be realized in different ways:

- *Industry consortia.* Incumbent producers and distributors form an industry consortium, thereby offering a one-stop shop for small and medium-sized customers. The consortium partners set up an independent entity which acts as a neutral marketplace, connecting buyers and sellers for a transaction fee. The new entity tries to incorporate all the customer's needs to build a real vertical e-network, taking new partners and services on board as appropriate. Its value will initially be based on the efficiencies of the transaction platform, but as those services commoditize, it will need to create additional value quickly through transparency and unbundling of services, as well as generating value from information and new services. Especially in the beginning, the new entity will depend on good relationships between supplier-founders to obtain support from marketing and sales in switching customers to the platform. The incumbent needs to secure its share of the variable equity (equity distributed according to the volume or sales channeled through the platform as an incentive to provide liquidity) of the new entity, and also needs to build its skills to market and differentiate itself successfully via the platform (major industry consortia are considering this). Attackers with in-depth expertise in setting up and operating vertical e-platforms and guaranteed neutrality may be appropriate partners for such an incumbent-owned industry consortium.
- *E-distributors.* Alternatively, an incumbent player can develop its own e-distributor platform which uses its superior value proposition to capture the customer interface and leverages this by selling products from different suppliers. Unlike e-marketplaces, e-distributors – such as global distributor GE Polymerland, for example – are not neutral. They own the goods and price them, often bundled together with other services into a unique package. There are several strategic routes to building an e-distributor. One option would be for a company to serve all small to medium-sized customers directly via its proprietary extranet, also offering products from competitors. A company can also build its own proprietary extranet and use it to e-embrace its existing distributors, enabling them to offer e-service and gaining stronger access to its end customers. Alternatively, it may e-enable selected large distributors (preferably controlled via equity stakes. GE Polymerland, for instance, is a wholly owned subsidiary of GE Plastics).

In the end, it seems most likely that only these two types of industry networks will survive in this segment, and that the e-distributors may possibly have the advantage in some segments. The individual supplier extranet will probably not succeed in this segment of high supply chain complexity because of the sheer number of websites a customer would have to handle in order to achieve transparency. The decision on whether to build an industry consortium or an e-distributor, in turn, depends on the size of the industry segment. With its higher complexity and the associated substantial upfront investment cost, an industry consortium will only be profitable in large segments. The high fragmentation of the chemical industry on the product level and the limited synergies between product groups restrict the size of homogeneous markets, however, and therefore favor the formation of e-distributors over industry consortia in some segments.

E-exchanges. The value proposition of an e-exchange, increasing supply and demand for well-defined products, implies that its natural owner will probably be an attacker company, on the grounds of its perceived neutrality. In addition, an attacker could provide additional skills, for example, online product testing and special logistics services, which do not necessarily belong to the core competence of chemical companies. Competition could arise from the industry consortia, which may also build these additional skills. However, their inability to exploit their key advantage of closeness to the customer here weakens their case. A company with significant sales volume in the industry segments most likely to be served by e-exchanges should actively invest in an appropriate attacker company in order to participate in the value created here and maintain control of its customer interface. However, for most chemical companies the actual sales volume generated by case by case supplier-customer relationships is quite small, so that a minority stake is probably the best way to participate in such a venture.

E-trading platforms. Because of the neutrality they offer to suppliers and customers, the natural owners of the e-trading platforms will be independent companies. However, any intermediary will depend on the support of major industry players on the supplier or customer side. The independent entity could, like Enron, also provide special financial skills outside the core competence of chemical companies. Chemical companies would be well advised to apply the same equity-holding approach to e-trading platforms as described above for e-exchanges.

Depending on the customer portfolio, more than one e-channel will probably have to be developed to serve the existing customer base. For example, extranets and industry networks will be the preferred e-channels for most industry segments served by the polyolefine industry (Fig. 7.3). One example of this strategy would be the injection-molding industry, where five large plastics manufacturers are building their own proprietary extranets, using the e-distributor GE Polymerland to some extent, and at the same time have founded the industry consortium Omnexus. In addition, most of the companies have taken a minority stake in e-trading platforms and e-exchanges like CheMatch and ChemConnect. Eventually, the customer will decide which of these e-channels survives.

7.4
How to Overcome Transitional Challenges

E-enabling the sales, marketing and supply chain processes will prove a major challenge for the incumbent chemical companies. The winners will excel in this transition by managing channel conflicts carefully during the transition phase, installing and maintaining one global customer database to ensure economies of scale and transparency, putting consistent business process rules in place to avoid arbitrage, and successfully moving customers online. They need to back up these moves by setting up a dedicated project organization to build the e-skills needed and to support critical change management.

Manage channel conflicts: The challenge here will be to build the online channels on top of the existing offline ones and integrate them seamlessly and quickly into one customer interface with several contact points. It is crucial to avoid any conflicts of interest between the sales force in promoting offline versus online solutions. The sales staff need to see e-channels as a tool to make them more effective and successful in interacting with their customers by eliminating a great deal of the highly repetitive routine work. Therefore, it is critically important to implement training programs and internal public relations work to ensure that the sales force understands the e-channels and their value proposition. Variable compensation should also be linked to sales volume and profitability, regardless of the media used for the transaction. Additional incentives should be created based on the penetration of online sales. In the case of Office Depot, a large distributor of office supplies, 50 percent of the sales representatives' annual bonus depends on their customers' use of the online channel.

In parallel to e-enabling their direct channels, companies typically have to manage the even more critical potential conflict with indirect channels. As a basic principle, the e-enabling of the distribution channel should be positioned as an opportunity to increase value to the customer rather than as a threat of reducing cost and margins, or eliminating the role of the distributor altogether. There are basically three approaches to introducing e-commerce for the indirect channel. These have to be decided up front, depending on the way in which distributors add value and the importance of the customer segment targeted:

- If a broad distributor base is important in delivering additional services to the customers, distributors should be e-embraced by providing software for distributors and their customers. On the one hand, this strengthens the distributors' customer front-end, and on the other hand helps the producer to improve its understanding of the customer base and its needs. This strategy is actively pursued by BASF Coatings in the refinishing market, by developing e-commerce solutions providing the recipes for millions of automotive colors on demand, and by DuPont, which provides applications which actually enable the company and its distributors to help customers in managing their paint shops.
- If the outlook indicates that e-enabling will lead to a rapid consolidation of the distributor landscape because of the investments involved and the ability of ag-

gressive players to consolidate, selected large distributors should be supported by providing them with superior e-service offerings to increase their competitiveness. The producer would ideally take a major equity stake in these large distributors, to secure control over the customer interface (as in the GE Plastics/GE Polymerland example).

- If the distributor does not play an important role in terms of value-added and customer access, a direct disintermediation approach may be advisable. Such a strategy is unlikely to lead to complete elimination of the distributor, since chemical producers will not want to build capabilities specifically for the complex supply chain management of small customers. However, the distributor's role will change more to that of a logistics service provider. The approach of some of the consortia remains to be seen here.

Install a global customer database: One critical organizational process is the management of all customer and transaction information, which is essential for capitalizing to the full on the increased availability and manageability of information. Therefore, successful players have installed a common customer database that is used via the intranet by the direct sales function, call center personnel, and the distributor, providing a consolidated view of all relevant customer data: master data, sales volume, lead activities, service levels, and so on. This also allows strategic marketing to understand and manage the value of the customer portfolio much more effectively. One state-of-the-art example here is Yellow Freight, which has integrated all relevant customer data into one web-based database.

Develop consistent business process rules: Based on this customer database, companies should develop consistent business process rules to handle increasing market transparency. In some segments of the chemical industry, historically grown inconsistent pricing can lead to substantial profit losses. Therefore, companies have to develop consistent value-based pricing rules to enable them to respond rapidly to customers' requests for quotations. In order to avoid customer arbitrage, that is, customers buying products on an e-marketplace and at the same time using value-added services directly from the supplier, different service levels have to be defined for each customer segment. Tight control of the customer registration process defining the e-commerce access levels allows for customer-specific service offerings.

Actively move customers online: Once the ground has been prepared, one of the biggest challenges is to actually move and retain customers online. To facilitate this process, a task force consisting of sales people with e-commerce knowledge should visit customers to conduct e-training courses on site and support the sales force in communicating the value. In many cases, suppliers will share some of their efficiency gains with their customers to incentivize the transition process. Once management is convinced of the value of the e-channel, every employee using it has to be trained, sometimes repeatedly, in order to accomplish the switch from offline to online sales successfully. According to Office Depot, the process of switching a customer completely to online dealings can take up to eighteen months. This task force will also have the role of collecting valuable customer feedback to further improve the e-commerce service offering. In addition, specialized help lines – sometime enabling escorted surfing – also support customers in working with the e-channel, with the aim of sustaining customer satisfaction.

To build new e-skills and ensure that the process optimization gains are actually realized, the e-enabling company will need a strong, dedicated project organization. After the transitional phase, this organization should be reintegrated into the company or business unit to enable consistent and synergistic multichannel management. The project organization for a business unit should be headed by an e-commerce manager, who is responsible for developing the e-commerce strategy together with the senior management of the business unit. The e-commerce manager should be supported by dedicated personnel: a content manager, a business process manager, an e-trainer and various experts covering IT-related topics such as web design, database structuring and master data consolidation. In most cases, only the e-commerce manager and the content manager actually have to be part of the business unit organization. They play the key role in transforming the organization into a web-enabled one, and have to be supported from within the business unit, with strong backing from the unit's manager. Corporate or divisional support departments to capture synergies across business units can provide the other e-commerce support functions.

When this project organization has been set up, chemical companies have to ensure that a consistent corporate IT platform is developed which can then be leveraged. This is particularly critical for e-businesses close to the core business, in order to capitalize on the synergies between existing business units. Here, the corporate IT department should serve as an internal e-commerce application service provider, although the development and hosting of the e-commerce solutions may be partly outsourced to a third party provider. Examples of such corporate IT platforms are solutions such as *MyAccount@Dow*, *MyAccount solution at BASF*, etc. E-commerce businesses that are not related to the core business of the chemical company can be set up as an independent company with no constraints on the IT side.

To initiate the required change management process across the entire corporation, finally, top management has to set a clear e-commerce vision that captures the commitment of the entire company. CEOs of some leading chemical companies have already stated that 30–50 percent of sales will be conducted online in 2003: this manifestly communicates the urgent need for the business units to develop a clear pathway toward a viable online business. In addition, many companies have appointed a global e-commerce leader who reports directly to the board of directors, or a separate corporate e-commerce department to catalyze internal knowledge development across all business units and regions, to provide preferred access to partners, and to steer investments into startups which are potential partners for the company.

That e-commerce represents an exciting opportunity for chemical companies cannot be denied. Equally undeniable, however, is the fact that they will have to undertake a major effort to transform their cultures to accept and then leverage the potential benefits of more efficient, transparent, and customer-specific e-business.

8
The Alchemy of Leveraged Buyouts
G. Sam Samdani, Paul Butler, and Rob McNish

Financial restructuring is all about leveraging the power of financial innovations to manage the risks and returns of investments better. Leveraged buyout (LBO) is one such innovation that uses a high level of debt in the capital structure of an acquisition, triggering fundamental changes in the way the acquired business is managed to create value. This LBO approach to value creation is gaining unprecedented popularity in the chemicals sector, especially in Europe, following its emergence in the USA in the 1980s. Over the last two years, a number of multi-billion-dollar acquisitions of chemical businesses by LBO firms have been in the spotlight. Cases in point are Laporte's selected specialty chemicals businesses bought by KKR (for USD 1.18 billion), Aventis' industrial gases business bought by Allianz and Goldman Sachs (USD 1.6 billion), Shell's epoxy resins business bought by Apollo (around USD 1.1 billion), Ciba's epoxy resins business bought by Morgan Grenfell (USD 1.14 billion), ICI's olefines, aromatics, polyurethanes and titanium dioxide businesses bought by Huntsman (USD 3.8 billion), and Zeneca's specialty chemicals businesses bought by Cinven and Investcorp (USD 2.1 billion).

What is driving this mushrooming of interest in chemical deals? Is it supply or demand driven, or both? Do LBO firms create more value than publicly traded chemical companies? If so, how do they do it? Perhaps more important, how could chemical corporations apply the lessons of the LBO firms to replicate their apparent success? In the same vein, as their own business model matures, what new routes can the traditional LBO firms take to improve their performance in the chemicals arena? Finally, what might the future hold for the competitive as well as collaborative interaction of chemical companies and LBO firms? What follows is a synthesis of findings based on a systematic review of our database of around 200 chemical LBO transactions completed since 1980, detailed financial analysis of several LBO deals, and interviews with dozens of LBO practitioners and managers with experience in buying, selling and operating chemical LBOs.

8.1
The Saga of the Deals

The successful chemical LBO practitioners are an intriguing bunch. Other than a few anecdotal stories, not much is known as to what makes them tick. Some of them, for example, Gordon Cain and Jon Huntsman, have become legendary by executing a series of deals and making themselves and many others around them very rich in the process. They could easily pass for the perfect hosts of a new game: "Who Wants to Be a Millionaire in Chemicals?"

To see how this game works in its simplest form, imagine an all-equity business entity that is bought for USD 1 billion. Let us assume that, before the acquisition, this business generated USD 100 million in cash flow, just enough to give shareholders a 10 percent return, and that the acquisition is financed with USD 900 million in debt and USD 100 million in equity. It is not unrealistic to suppose that, through aggressive operational improvements, superior asset utilization, and careful capital investment, the business is able to double its cash flow from USD 100 to USD 200 million per year, without either increasing or decreasing the value of its assets. By using this USD 200 million in cash flow strictly for debt service, this business can pay off the USD 900 million of debt (at an interest rate of 10 percent) in about six years. At the end of that period, the business would still be worth USD 1 billion, but it would now be all equity. In other words, the original USD 100 million equity investment has been converted into one worth USD 1 billion, for a compounded annual rate of return of 47 percent. This is how financial leverage concentrates ownership in fewer hands and substantially amplifies the returns to the new owners.

In his autobiography, *Everybody Wins: A Life in Free Enterprise*, Cain (1997) talks about how ownership restructuring can benefit everyone, and how the free market economy provides unparalleled opportunities for all. A chemical engineer and industry veteran, he did his first LBO in 1983 at the age of 71 through the acquisition of a large piece of Conoco Chemicals from DuPont. Subsequently, he acquired several commodity businesses from Monsanto and ICI and turned them around. He is especially proud of the fact that, even in the tumultuous 1980s, a "gentleman" managed to finish on top and generate millions of dollars in the process.

A former US Navy officer and aide to President Nixon, Huntsman started out in 1970 with a foam-packaging business and bought his first chemical plant from Shell in 1982. He has since built up the world's largest privately held chemical company through a series of LBO deals, for example, Texaco's chemical operations in 1994–97 and a huge chunk of ICI's chemical businesses in 1999.

We believe that the visible successes of Cain, Huntsman, and a few other industry insiders (e.g., George Harris, Hal Sorgenti) have in recent years attracted a lot of imitators, especially financial buyers, into the chemicals sector (Fig. 8.1). Industry observers point out that given the low public valuations of chemical assets and the unprecedented levels of uninvested funds available today (shown in Fig. 8.2), chemical businesses make ideal LBO targets. Their logic is that the basic industrial sectors, such as chemicals, have reasonably predictable cash flows, unlike the all-too-many high-tech and Internet startups, for example, boasting stratospheric valuations with negative/unpredictable cash flows. This is the demand-side explanation for the recent explosion of chemical LBO deals (Fig. 8.3).

The supply-side story is equally convincing: large chemical conglomerates have found themselves under increasing pressure to go down the path of strategic focus by shedding various non-core assets. The big increase in LBO deals in Europe in recent years has been driven by such portfolio restructuring by the chemicals and energy conglomerates who started to divest their small and specialty chemicals assets in order to focus on their core businesses, causing a shift in the deal flow away from commodities to include specialties, polymers and fibers as well (Fig. 8.4). In this environment, LBO firms of all shapes and sizes entered the market and have become very competitive with chemical companies, especially in smaller transactions. We believe that competition for deals is likely to intensify, leading to differentiation and specialization among the LBO firms. The good

Fig. 8.1 Fragmentation of LBO players in chemicals, 1980–2000
* Based on deals completed as of October 2000; Source: McKinsey analysis

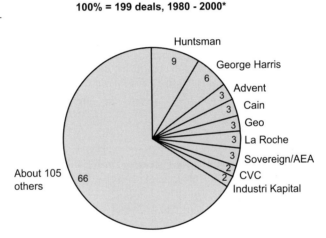

Fig. 8.2 Unprecedented levels of uninvested funds, 1992–1999 Source: VentureXpert; interviews; McKinsey analysis

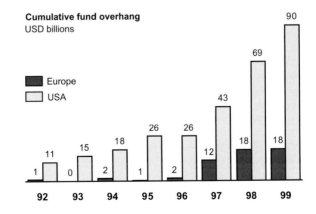

USD billions

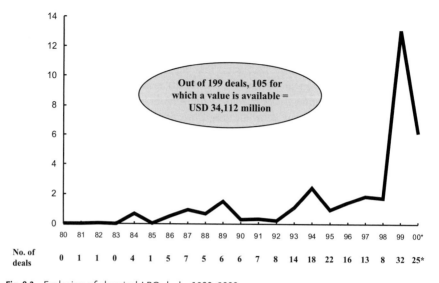

Fig. 8.3 Explosion of chemical LBO deals, 1980–2000
* Based on deals completed as of October 2000; Source: McKinsey analysis

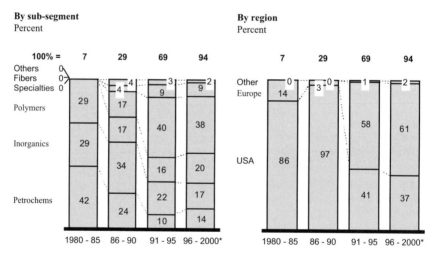

Fig. 8.4 Deals shift from commodities to specialties and from USA to Europe, 1980–2000
* Based on deals completed as of October 2000; Source: McKinsey analysis

news is that deal flow has become extremely varied over the last couple of years, allowing the LBO players to be picky as well.

8.2
What is Attractive about Chemical LBOs

The high level of interest paid by LBO firms to chemicals appears to be justified by the returns. Our research indicates that, over a 20-year time frame (1980–1999), LBO investments in chemicals have clearly outperformed both the S&P 500 and the all-sector LBOs, while over the same period publicly traded chemical companies have slightly underperformed the S&P 500 (Fig. 8.5). This cannot be explained by leverage alone, since LBO investments in chemicals have outperformed publicly traded chemical companies even on a leverage-adjusted basis (Fig. 8.6). Over the last five years, however, the bull market managed to push the performance of the S&P 500 beyond what LBO investors could achieve.

We believe that the chemicals sector has characteristics with broad appeal to LBO players. For example, it is rapidly maturing, with thousands of poorly managed companies that have great potential for performance improvement through cost reduction alone. Many segments (e.g., adhesives, specialty coatings, biocides, food additives, industrial/institutional cleaners, etc.) are highly fragmented and are prime candidates for roll-up by LBO players. Massive restructuring, especially in Europe, will ensure high deal flow for many years to come. In addition, many of the family-owned businesses in Europe, especially in Germany, offer the opportunity to negotiate exclusive deals and avoid the bidding war at competitive auctions.

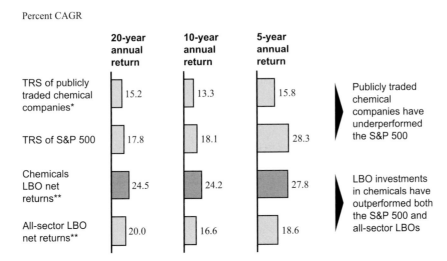

Fig. 8.5 Comparison of reported returns to investors, 1980–1999
* Composite of US publicly traded chemical companies' TRS (capital appreciation plus dividends); ** Includes leveraged buyout fund investments in later stages (e.g., MBO, LBO, mezzanine). Does not include venture capital (e.g., no seed, startup, pre-IPO). Returns are total return to external investors net of all fees as of 12.31.99; Source: Venture Economics, McKinsey analysis

Percent CAGR

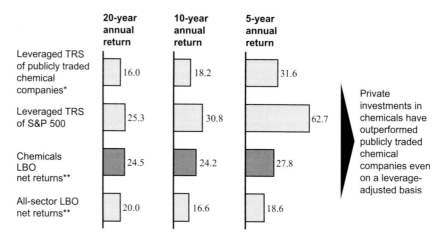

	20-year annual return	10-year annual return	5-year annual return	
Leveraged TRS of publicly traded chemical companies*	16.0	18.2	31.6	
Leveraged TRS of S&P 500	25.3	30.8	62.7	Private investments in chemicals have outperformed publicly traded chemical companies even on a leverage-adjusted basis
Chemicals LBO net returns**	24.5	24.2	27.8	
All-sector LBO net returns**	20.0	16.6	18.6	

Fig. 8.6 LBO investment returns relative to risk-adjusted benchmarks, 1980–1999 * Composite of US publicly traded chemical companies' TRS (capital appreciation plus dividends); ** Includes leveraged buyout fund investments in later stages (e.g., MBO, LBO, mezzanine). Does not include venture capital (e.g., no seed, startup, pre-IPO). Returns are total return to external investors net of all fees as of 12. 31. 99; Source: Venture Economics, McKinsey analysis

8.3
How Value is Created in Chemical LBOs

The popular press on LBOs has been less than flattering. For example, *Barbarians at the Gate: The Fall of RJR Nabisco* by Bryan Burrough and John Helyar (1990) depicts the LBO players as amoral creatures driven by greed and false glory. LBOs have also been criticized for making it possible for a select few to get rich at the expense of many who get nothing at all. However, Gordon Cain's book, *Everybody Wins*, offers a different picture altogether.

In the chemical sector, the received wisdom is that in their relentless focus on making money, LBO practitioners resort to "slash and burn" tactics, compromising environmental health and safety. They are sometimes referred to as the "bottom fishers" looking for a "quick flip". In addition, they are often described as financial magicians who turn solid balance sheets into smoke and mirrors.

Our analysis and interviews suggest that the reality is different from the above myths about chemical LBOs. For example, the facts are quite contrary to the belief that health, safety and environment (HSE) issues in chemical plants are compromised under LBO firm ownership. One manager spoke of a radical change under an LBO owner from a situation where the HSE record was so poor it was impossible to get insurance to one where "the company's reception area is plastered with safety awards." Another LBO, Victrex, won awards from the UK's Royal Society for Prevention of Accidents in each of the first two years following the buyout. Of course, such a policy also makes sound commercial sense, since it is far easier to exit from a business with a good safety record.

The trends in buy-hold-sell cycles of chemical LBOs do not support the "quick flip artists" myth either. In fact, the holding times for chemical LBOs are much longer than is generally believed. Our database of 200 chemical LBOs covering deals since 1980 shows that fewer than one third of them exit within five years. More pointedly, our analysis suggests that LBO players create more value in operating the assets than in trading them.

Since financial data for most individual LBOs are not reported publicly, we analyzed nine chemical LBOs that subsequently exited via initial public offerings (IPO) and for which a financial history is available. The analysis shows that about two thirds of the value captured by the LBO firms is created during the holding/operating period, and only about one third during the buying and selling processes (Fig. 8.7). LBO firms are tough operators committed to taking out unnecessary costs and improving capital productivity. Our interviewees cited many examples of post-LBO actions taken to improve operating performance: corporate centers downsized from 300 to 180; 40 percent of white collar positions eliminated; fixed costs reduced by 35 percent in 12 months; R&D cut back to focus on projects with a maximum payback period of three years; prices increased successfully by 13 percent, and so on.

Finally, smoke-and-mirrors financial engineering, especially the use of debt to replace equity, does not create a lot of value in and of itself. It turns out that the cost of capital is more or less independent of leverage, since the tax advantage of a high level of debt is almost entirely offset by the higher cost of that debt. How-

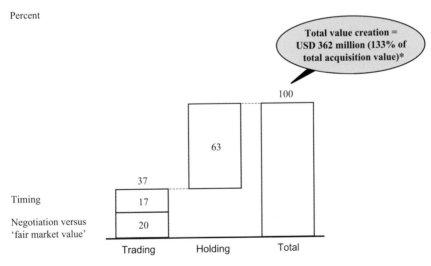

Fig. 8.7 Sources of value creation
* Sample of 9 deals for which required financial details are available (Sterling Chemical; Victrex; UCAR; ChiRex; Inspec; Vista; Arcadian; Brunner Mond; ISP); Source: McKinsey analysis

ever, the high leverage, together with senior managers' personal "pain" equity, provides the incentive to jumpstart business performance. Having to service and repay debt forces managers to concentrate on generating cash, while putting personal funds – often raised by borrowing from the bank or by remortgaging the family home – into equity provides the motivation to deliver results. No less motivating is the expectation of a big earn-out.

Our interviews with LBO managers and investors confirm that they create value in operations by aligning management and shareholder interests through strong personal incentives linked to EBITDA or cashflow. Indeed, the difference between the incentive packages for executives in publicly traded chemical companies and those in LBOs is remarkable (Fig. 8.8). The major disparity is the link between effort and reward. Managers with experience in both environments point to the weakness of the link between what they actually did in a large corporation and their non-salary rewards. Extra effort resulted in no greater payback, since the value of stock options, bonuses, phantom shares or similar devices was subject to many forces outside their control. In an LBO, by contrast, additional effort gets translated directly into bonuses and – after some delay – an equity cash-out.

	Chemical corporations	**LBO acquisitions**
Leverage (typical)	20%	70%
KPIs	RoNA; RoACE; CFROI	EBITDA; cash flow
Equity stake (typical)	Very small	1-2%
Type of investment	Options	Direct; up front
Bonus (typical)	Up to 25% of salary	Up to 60% of salary
Incentives' link to performance	Weak	Very strong

Fig. 8.8 Leverage and incentives for managers
Source: McKinsey analysis

8.4
Infecting Chemical Corporations with the LBO Virus

The LBO approach offers a fresh alternative to the traditional modus operandi for chemical corporations. Managers we interviewed talked of emotional highs and the excitement of "LBO fever" when released from corporate bureaucracy. Others spoke of a massive cultural change, an acceptance that there are "no excuses," and a sense that failure is not an option.

We believe chemical corporations could unlock significant value by applying the lessons of LBO firms to many businesses in their portfolio. For most multibusiness chemical corporations, the list of suitable candidates could represent a sizeable share of their total revenues. Characteristics of the LBO-ready businesses include: chronic underperformers unable to achieve breakthrough performance inside a large corporation (e.g., the chloralkali chain), businesses where demand growth is static or negative and where the focus has shifted to improving capital productivity (e.g., fertilizers), business units where product/process and market maturity has tilted the balance away from innovation and toward harvest (e.g., fibers), businesses where cashflows are strong and stable enough to support high leverage (e.g., cellulose acetate), and businesses where lack of portfolio synergies allows effective isolation of individual business units (i.e., most chemical company portfolios).

However, applying the LBO approach to such diverse candidates is not necessarily straightforward. There are five broad ways in which chemical corporations could exploit the lessons from LBO practitioners. These range from the low-risk objective of improving M&A and disposal performance to operating like a financial holding company to leveraged recapitalization of the whole company. In between, companies can consider carrying out do-it-yourself or collaborative "internal LBOs" of selected businesses.

Of course, the approach taken will vary from company to company, and will depend on the nature of the portfolio and the attitudes of the CEO and CFO toward risk and unfamiliar financial structures. The capabilities required and the reactions of stock markets are also likely to be very different for each option.

8.4.1
Maximizing M&A Performance and Divestment Value

Historically, chemical corporations have overpaid for acquisitions by overestimating the synergies. In addition, they are often not very good at disposing of their underperforming businesses. We believe that chemical companies could significantly improve their M&A performance by wholeheartedly adopting the best-of-breed LBO firm mindset. Of course, companies have to overcome strong emotional feelings about the current portfolio and potential acquisitions and other organizational/behavioral issues, e.g., slow decision making processes and managers' traditional discomfort with M&A as a way of creating shareholder value. However, we believe it can be done through consensus building and performance tracking (e.g., treating an M&A department as an internal profit center very much like LBO firms do).

In addition, selling non-core businesses to LBO firms is low risk and in many cases is preferable to hanging on to underperforming assets indefinitely. Given that the LBO market is now sufficiently competitive, chemical corporations should seize the opportunity to capture "fair" prices for their unwanted businesses. To ensure that the final price is the maximum achievable, they should ensure competitive auction until the last minute, and maintain the involvement of senior managers to increase the buyer's confidence in the quality of the management. They could also maintain a small equity stake (say 10 or 20 percent, possibly with a seat on the board) for some probable financial upside and the privilege of learning first hand how LBO firms achieve their turnaround performance.

8.4.2
Operating Like a Financial Holding Company

Chemical companies could behave more like financial holding companies to bring the "LBO magic" to their business units (BUs). We believe that internal synergies in chemical companies are quite often overestimated, the cost of the corporate center may far exceed its value, and that differences between BUs are greater than the similarities across them. We also believe that this organizational structure could unleash the impact of balance sheet restructuring and strict financial discipline within the BUs. The organizational features would include a small board, fewer interfaces between the corporate center and operating units, financial structures and key performance indicators (KPIs) tailored to business specifics, managers who are strongly incentivized to perform, and contractual agreements to ensure that unsuccessful managers are not simply moved from post to post.

8.4.3
Implementing Do-it-Yourself 'Internal LBOs'

The purpose of an internal LBO is to achieve the same level of value creation as an LBO firm would aspire to, but to capture the benefit for the company's own shareholders. The concept sounds deceptively simple but the small number of examples of such internal LBOs is a testament to the difficulties that corporations have with contemplating, planning and implementing an internal LBO. There are at least three possible models of non-recourse debt-financed recapitalization that chemical corporations could apply at the business unit level to create extraordinary value. These are internal carveouts, leveraged partial public offerings, and management buyouts (MBOs). The guiding principles for all three models include high leverage (debt/equity ratio of three, for example), significant management incentives tied to equity purchased with personal funds, complete operational and strategic freedom for management, and acceptance by all parties of a set of possible exit routes similar to what might be expected in a real LBO.

In all these internal LBO/MBO models, the alchemy of leverage turns the key managers into owners of their businesses, who, in turn, end up converting mundane chemical businesses into "money machines." In addition to concentrating a significant equity stake in the hands of operational management, debt repayment obligations can, and often do, remove the temptation for over-investment or value-destroying acquisitions, and stimulate the sale of underperforming assets for which a better natural owner exists. Furthermore, debt refocuses attention on cash flow, which obviates the common concern with accounting-based reported profits and imposes strict financial discipline.

8.4.4
Harnessing the Catalytic Power of Collaboration

By contributing their well-honed deal-making and leverage-driven performance improvement skills, the best-of-breed LBO firms could be the catalysts or sparring partners that chemical companies need to get going. As an illustration, something like 51 percent of the equity may be held by the LBO firm, 39 percent by the chemical company, and 10 percent by operational managers (depending on the size of the business). The LBO firm negotiates the deals and arranges non-recourse debt financing for high leverage. There may be put/call options in place for exit in four to five years, for example.

In this partnership form, the business enjoys true separation and independence from the corporate center. However, the downside is that considerable value is transferred to the LBO firm.

8.4.5
Executing Leveraged Recapitalization

Leveraged recapitalization of an entire company is a significant step away from internal LBOs, but it still contains the essential elements of the LBO approach, that is, increased financial leverage plus equity incentives for senior managers to significantly improve performance. The difference between a leveraged recapitalization and taking the whole company private is that a recapitalized company retains its public shareholders, has stock that is quoted on a public stock exchange, and publishes financial data in the same way as a publicly traded company does. However, such a recapitalization increases a company's leverage and concentrates more of the equity in the hands of senior management (and possibly other employees), thus strengthening the incentive to maximize performance.

Two chemical companies executed leveraged recapitalizations in the 1980s. One was Union Carbide in 1985 when it faced a hostile bid from GAF. Carbide sold its Eveready Battery business and used the proceeds to buy back 55 percent of its shares at a price that GAF could not match. The other leveraged recapitalization was that of FMC in 1986, when it decided to return a large portion of its cash to public shareholders in a kind of "public LBO." In the transaction, FMC's senior managers increased their equity stake in the company from 19 to 41 percent.

8.5
New Roles for LBO Players

Despite the flurry of LBO activities in chemicals today, or perhaps because of it, there are some clouds on the LBO horizon. Thanks to the growing sophistication of chemical companies in valuing their businesses, the negotiating power between buyers and sellers is balancing out, and the nature of the deals is becoming increasingly complex. In this environment, success requires creativity and innovative approaches to deal structuring and execution. The current deal boom in Europe, expected soon to be followed by a similar one in Asia and Latin America, requires the ability to acquire, operate, and sell chemical assets run under different tax regimes and accounting rules along with varying environmental and safety regulations.

The recent rise in LBO activities has also created a little known phenomenon: the largest number of deals that have yet to exit. Our research indicates that only one third of LBO deals closed over the last twenty years have exited, most sold to trade buyers, with IPOs becoming an unlikely exit route (Fig. 8.9). There are indications from our interviews that some LBO owners are now becoming impatient with the increasing difficulty of achieving the kind of exit they had once hoped for. There is an increasing sense that the real test for the more recent LBO players in chemicals will come when they seek their exits. Some industry observers indicate there is now reason for caution about the rate at which the chemical LBO market will continue to grow.

Given the impending exit challenges and the growing diversity and size of chemical deals, LBO players have to fundamentally rethink their roles and busi-

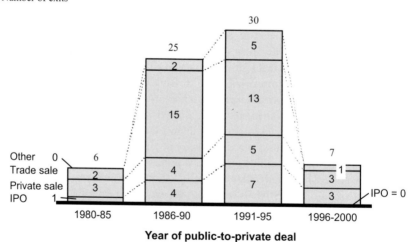

Type of exit over time, 1980-2000*
Number of exits

Fig. 8.9 Exit by IPO becoming increasingly difficult, 1980–2000
* Based on deals completed as of October 2000; Source: McKinsey analysis

ness models going forward. We propose several new roles and business models for those who want to become distinctive and more successful.

8.5.1
Portfolio Optimizer

Many diversified industrial companies added chemical assets to their portfolio over the years without a clear and coherent strategy for creating superior returns for shareholders. Now that most of them are beginning to focus on their "core" businesses but are reluctant to dispose of the mixed bag of chemical assets in a piecemeal fashion (e.g., Zeneca's decision to sell all its specialty chemicals businesses in a single transaction), we are seeing the emergence of a new breed of LBO firms (e.g., Cinven, Investcorp, AEA Investors) who are carving out a role for themselves as portfolio optimizers. We believe that these LBO players could create significant value by bringing rapid decision making and relentless focus on cash flow growth to chemical companies while streamlining their portfolios prior to exit.

8.5.2
Aggressive Consolidator

Some LBO players specifically seek a restructuring role within certain sectors of the chemical industry. Huntsman is one of the best examples, with the creation of a polystyrene business from a variety of acquisitions in the USA and Europe (subsequently sold to NOVA Corp.). George Harris is another, with a global salt and soda ash business that was built up from a number of acquisitions and then sold as a package to IMC Global.

A large number of specialty chemicals businesses currently owned by chemical/energy conglomerates, especially in Europe, are subscale and could benefit from segment-level consolidation. Examples of such fragmented segments include adhesives, sealants, specialty coatings, food additives, cosmetic chemicals, biocides, and many other specialty and semi-specialty segments. LBO firms are better positioned than traditional chemical corporations to rapidly consolidate these fragmented/undermanaged segments and capture the most value.

8.5.3
Spanning the Alliance Spectrum

We believe that solo acquisitions will become increasingly difficult and risky because of the large size and complexity of many chemical deals. Allying with chemical companies and other LBO firms with complementary skills and resources offers the opportunity to structure innovative acquisition, operation and exit deals.

Cinven partnering with Investcorp to acquire Zeneca's specialty chemicals businesses (Avecia) is one such example. Another is the joint venture structured by Lehman Brothers with Hercules to acquire the Kelco biogums business from

Monsanto, with Hercules contributing its own food gums business for 28 percent of the equity and Lehman retaining the remaining 72 percent. The alliance spectrum for these kinds of deals may include contractual agreements (e.g., for shared services and infrastructure), minority stakes (e.g., operational "keiretsu" that leverages privileged relationships among a network of chemical suppliers and customers), joint ventures (usually 50/50 contribution of assets, capabilities, technology, etc., to a separately managed entity), and partnerships (shared roles and responsibilities as partners in a single combined entity).

8.5.4
LBUs: Moving beyond the Conventional LBOs

Given that the opportunities for value creation through traditional "buy low, sell high" approaches are going to be limited, LBO firms need to build businesses from growth platforms to create leveraged buildups (LBUs). Key success factors include deep industry expertise and market insight/foresight. With Bob Covalt (bringing deep industry knowledge from his days at Morton International) and other industry veterans on the team, AEA Investors appears to be heading in this direction with its follow-on acquisitions to build a growth story for its Sovereign Specialty Chemicals in niche adhesives, sealants and coatings segments.

8.6
Steps toward a Natural Convergence

Chemical corporations are usually led by engineers and chemists eminently qualified to manage businesses with manufacturing plants full of mysterious equipment and hazardous materials. They are the modern corporate alchemists who convert a handful of building block molecules into thousands of synthesized and formulated chemical products. These engineers and chemists are understandably mistrustful of investment bankers running chemical plants. Furthermore, the traditional business models of multibusiness chemical corporations – focused on making products by the tons – and LBO practitioners – focused on making money by the millions – seem to have evolved in parallel universes. However, the LBO players are now emerging as formidable competitors in many segments of the industry, often outbidding publicly traded chemical companies for acquisitions. In addition, they are winning the competition for capital and the war for management talent. Therefore, chemical corporations cannot treat the LBO players as aliens from a different universe anymore. They will have to benchmark their performance against these new competitors, and even be prepared to embrace them as partners.

With the continuing restructuring of the chemicals industry, the deal flow will continue to create opportunities for those with distinctive insight and foresight. Players with superior execution skills will continue to be handsomely rewarded. As discussed above, given the historic outperformance of the LBO players in chemicals, chemical corporations may begin to take steps to replicate their success. Given the challenges that the chemical LBO players are likely to face going forward, however, they may also start taking the necessary steps to differentiate themselves through more innovative approaches to deals, including collaborations with chemical corporations.

Chemical corporations and LBO firms may move toward a "best of breed" model in which the complementary skills and resources of these two types of players are effectively leveraged to benefit both. In this natural, not necessarily inevitable, convergence, those who move first to reach across their respective cognitive barriers for creative collaboration stand to gain the most.

9
Revitalizing Innovation
Wiebke Schlenzka and Jürgen Meffert

As the flow of new chemical molecules dries up, the chemical industry needs to look to other sources of innovation in addition to traditional chemical research, and also ensure that it captures the maximum value of each innovation. In order to do both successfully, chemical companies have to break down their traditional inward orientation, determine explicit strategies for innovation, and mimic the business patterns and mentality of successful venture capitalists and new startups to take advantage of outside resources. Among the attractive new technological sources of innovation that they should explore are biotechnology and e-commerce (see Chapters 6 and 7).

This chapter examines how chemical companies should develop new innovation strategies and some practical routes to their implementation.

9.1
The Innovation Challenge

Without much exaggeration, the last century could be called the century of chemical innovation. Plastics and other synthetically synthesized molecules and materials changed most industrial and consumer markets tremendously by systematically replacing natural products like paper, wood, or cotton and making completely new applications possible.

However, in the last two to three decades of the twentieth century the innovative force of the chemical industry slowed down. Far fewer really new molecules were developed and the change in properties compared with those of existing materials became incremental. In addition, key markets have become saturated and production costs are getting closer to core costs. In this environment, many chemical companies are in desperate need of innovation to generate the profitable growth which is a major driver of their stock value (see Chapter 3).

This seems paradoxical, coming at a time when the pace of knowledge development, accelerating at levels unimaginable a decade ago, has led to tremendous innovations in a very wide spread of industries. Gene technology, for example, has changed the pharmaceutical industry's approach to drug design and has already led to the development of drugs for diseases that could not be treated in the past. Epogen from Amgen or TNF from Genentech are examples of top selling biotech drugs. Today, there is no major pharmaceutical company which does not apply gene technology in drug development. In telecommunications, wireless and fast data transfer technologies have revolutionized the entire industry. Finally, the Internet has started to change the traditional value chains in retailing and trading.

However, small startup firms have cornered the markets here, while established companies are struggling to maintain their innovative edge, and the same applies to the chemical industry. Innovations in e-commerce and in biotechnology have started to change the game completely in some chemical businesses, such as agro-chemicals, but most of these originate with startups, and not with the big established chemical companies.

How can chemical companies regain innovativeness? To do this, they have to open up their mindsets and have to think in terms of innovations beyond molecules. They need to transform themselves into "incubators" that are open to the external community and that continuously generate ideas and develop them into commercially successful businesses. This requires a clear definition of the company's innovation strategy and the establishment of company-specific practical approaches to stimulating innovation.

9.2
Defining the Innovation Strategy

First and foremost, a successful innovation strategy should focus on the high impact – transformational and substantial – types of innovation. Having set this aspiration, chemical companies should then develop tailored strategies outlining the areas in which they will seek innovation and ways to overcome any current issues that block high impact innovations.

9.2.1
Focusing on High Impact Innovations

When people talk about innovation, they sometimes really mean invention. Inventions may yield a patent, but do not necessarily create economic or social value – they may never go beyond an interesting description or a first prototype. To become an innovation, a new development has to do much more: it has to have a tangible impact on companies and consumers.

Depending on their impact, we can distinguish three different types of innovation at the corporate level.

Transformational innovation. Here, completely new markets are created, and whole industry structures are changed fundamentally as a result. Think, for example, of the development of optical fiber with infinite transmission capacity, which made the old copper technology obsolete in long distance telecommunications, or the development of polyolefine plastics as packaging materials. This allowed sterile and long term conservation of many food products, making other packaging materials obsolete and also eliminating the need for local food production.

Substantial innovation. These are new generations of products, services, or processes that significantly change the balance of power between competitors in a given industry. Examples are the introduction of Plexiglas in the fifties/sixties and DNA purification in the eighties.

Incremental innovation. This is a natural part of the continuous improvement process in corporations – traditional product development efforts leading from one product to its next generation. Here we can cite various specialty chemicals, such as de-watering agents, coating materials, or colorants whose product characteristics such as stability, color, and solubility can be improved by further modification of the existing molecule or by mixing different substances.

Up to the 1970s, chemical companies were able to generate a wealth of transformational and substantial innovations through the development of new molecules by chemical R&D. However, the opportunities for identifying new molecules through classic chemical research have decreased significantly, and as a result many chemical companies are now focusing their research on incremental innovation: in other words, improving existing molecules, materials, and production processes. This move to incremental innovation is often not a deliberate strategy, but rather a natural consequence of the fact that the two high impact types of innovation are becoming more and more difficult to achieve with the traditional approach. In chemical companies today, these efforts account for up to 90 percent of a total R&D spend which is already low by comparison with other industries. For example, pharmaceutical companies spend between 10 and 20 percent of their total sales on R&D; chemical companies often spend less than 5 percent. As incremental innovations have much less impact, they provide poor fuel for corporate growth. This is supported by the fact that there appears to be no correlation between chemical companies' R&D spend and increases in sales (Fig. 9.1).

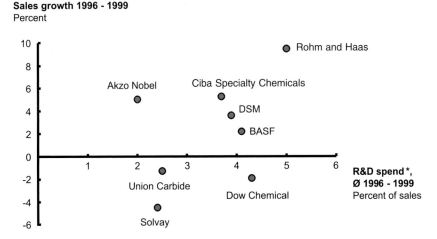

Sales growth 1996 - 1999
Percent

Fig. 9.1 Sales growth versus R&D spend
* Companies with chemicals making up more than 70% of total business; Source: Chemical & Engineering News issues, McKinsey analysis

Indexed venture capital committed to US venture funds*
(1980 = 100)

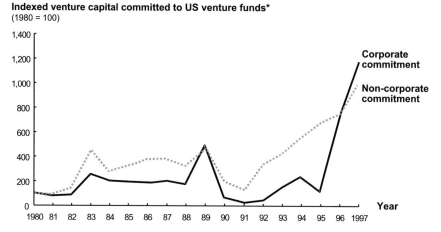

Fig. 9.2 Resurgence in corporate venturing
* Does not include direct investments in startups by corporations; Source: National Venture Capital Association, Annual Report 1997

Incremental innovation can be a very successful strategy for an individual company, but it can be life-threatening if an entire industry fails to seek long term sources of new growth. Chemical companies therefore need to raise their aspirations, and redirect their efforts back to transformational and substantial innovation. On the one hand, they should revamp their internal R&D management and focus it on these types of innovation, and on the other, they should open up to the world outside their own companies, as most new technology developments that can drive transformational and substantial innovation in the chemical industry are being developed by external startups or research communities: e-commerce, information technology, and biotechnology, for instance.

Of course, chemical companies did not rely solely on their own internal research in the past; important research alliances with top research institutions have formed part of their strategies for many years. Nevertheless, the traditional collaboration between big companies and research institutes is a much less successful model nowadays. The exponential increase in the availability of venture capital has made talented researchers start their own businesses instead of handing major lucrative ideas to established companies for a relatively small return (Fig. 9.2).

9.2.2
Defining the Specific Company Strategy

Once they have established their overriding innovation goal, chemical companies need to work out a strategy tailored to their more specific aims and requirements. The starting point should be a diagnosis of the company's core business to identify its overall competitive position, current growth prospects, and investors' growth expectations. Some companies – many pharmaceutical companies are a case in point – might find that investors have built much higher growth expectations into their current share price than they can achieve with their present efforts (see Chapter 2). In this case, they need to stimulate innovation to meet the investors' expectations. A more likely predicament for chemical companies is a low share price in relation to peers, indicating that investors do not think they are capable of substantial profitable growth. To increase their share price, the key is to build major valid options for growth quickly.

While investor expectations will delineate the size of the playing field, benchmarking against competitors will give the company more precise information about its strengths and weaknesses in innovation. Specifically, it should compare itself with its peers along the following dimensions:

Successful commercialization of innovations. Good indicators here are the number and sales of innovations achieved over the last few years and their share of total sales. Innovative companies find that recently launched products generate major revenues compared with the rest of their portfolios.

Type of innovations. As outlined above, strong innovators focus on transformational and substantial innovations. However, some companies may find that they are pursuing high-impact innovation without commercial success, which will alert them to the fact that there is a block to be sought elsewhere in the organization. In addition, they need to consider the area for innovation here: whether they are strong in in-house R&D, for example, or whether they have major strengths or gaps in other areas of opportunity such as effectiveness in services or other non-traditional parts of the value chain, or in the new technologies. Providing services, especially in the application of chemicals, is a strength that has become an increasingly important source of innovation for chemical companies (see Chapter 3).

Success in development. Measures of quality here are the break-off rates over the stages of development and the level of relative investments in break-off projects. Low break-off rates in the early stage, high ones near the end, and high investments in break-off projects could indicate a lack of rigor in monitoring the progress of R&D projects.

A clear picture of its relative competitive position in innovation and its strengths and weaknesses will help the company to pinpoint the most promising areas – for example, better nurturing of good internal ideas or aggressive use of a current strong market position to gain access to external ideas – and to design an action program for innovation. However, before developing such a program, it is important for management to understand the implications of such a move: it requires a high level of commitment of both capital and human resources over a long period of time. It also calls for new management approaches, and in most cases major organizational change.

9.3
Approaches to Stimulating Innovation

Many people think that innovation cannot be planned. It is, of course, true that some individuals are more creative than others, and that there is no way of training employees to have the spark of genius. Nevertheless, programs can be set up to ensure that the best ideas are identified and fostered, and the right environment can be created.

At the beginning of substantial or transformational innovation, both the customer segments and the eventual definition of the product are largely unknown. This makes it vital to "think startup", to build an entrepreneurial mentality and business management skills.

To grow new ideas into flourishing businesses, companies must define them as separate entities that can develop autonomously like startups in the open market. This ensures that they will get the market feedback they need in order to thrive – and also provide protection against corporate fears of cannibalism.

Some effective tools are available to make this transformation happen:

- Clearly structured idea generation processes to build the innovative and entrepreneurial spirit in house.
- Corporate venture funds to gain access to high potential ideas and deals from outside the company.
- Startup-type business structures and incentive systems with the right level of parent company support to grow the businesses successfully.

9.3.1
Clearly Structured Idea Generation Processes

Established industrial companies often have a built-in negative loop that blocks innovation. Slow or declining growth leads to cost cutting and downsizing. This, in turn, causes a skewed staff age profile, leads to recruiting difficulties, and culminates in a pessimistic culture. In such an environment, a non-entrepreneurial spirit, stifling new growth, comes as no surprise. How, then, can companies break out of this negative loop, inspire employees, and create innovative and profitable growth? Clearly defined processes for idea generation are one effective way to propagate ideas and build entrepreneurship. Some of the most important of these processes are "plug into the world" trips, killer idea workshops and – most importantly – business plan competitions.

Plug-into-the-world trips give a group of people from the company special exposure to external ideas, alternative approaches to problem solving, and new applications for existing platforms. Visits are arranged to universities, new ventures, venture capital companies, successful competitors, and customers, for example. This intensive exposure right across the marketplace can dramatically increase the likelihood of new ideas arising.

"Killer ideas" is a highly adaptable tool designed to stimulate the generation of ideas by posing a series of provocative questions in a workshop setting. The killer idea brainstorming process is designed to enhance both the quantity and quality of ideas generated as well as to drive forward the implementation of those ideas. The approach employs a rigorous set of evaluation criteria in a tree of structured questions to separate more substantial innovations from opportunities for mere incremental improvement. The structure makes sure that the group members really focus and think about what they are doing, and the stringent criteria ensure that these ideas are so good that they kill traditional thinking about the business (hence the name). This technique often brings forth ideas which have either genuinely never been previously considered, or which people have not dared to mention.

Business plan competitions are now a well-proven tool for surfacing ideas for new businesses in both a regional and a corporate context (Fig. 9.3). In one American life sciences company, for example, 1000 ideas were generated. These were turned into 100 business plans, of which 56 received seed funding. In a business plan competition, participants develop a fully-fledged business plan, usually in three phases from the initial description of the business idea to concrete details such as the appointment of the management team and a financial plan (Fig. 9.4). The owners of the most promising ideas are given coaching to further fine tune their business plans, some startup financing, and special independence in the parent company to develop their businesses.

Truly effective business plan contests are high-energy, high-profile, and highly professional affairs, generally sharing the following characteristics:

• Extensive media exposure and mobilization, with clear commitment from top management.

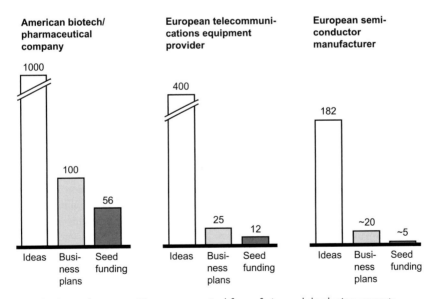

Fig. 9.3 Business plan competition – a proven tool for surfacing and developing concepts
Source: McKinsey

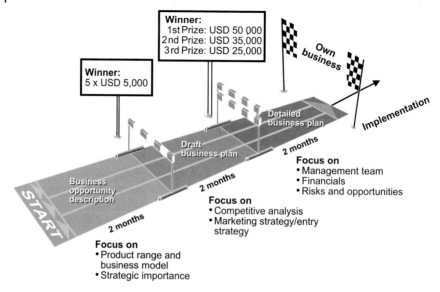

Fig. 9.4 Key phases of business plan competitions
Source: McKinsey

It is important to get enough people involved to create an innovative momentum: with internal and external press releases, town hall meetings where top managers state their commitment, special intranet pages, workshops with successful entrepreneurs from startups or non-competitive industries as guest speakers, attractive awards for the winners that get the whole organization talking, and internal hotlines.

- Substantial education and coaching.
  It takes coaching and training to get the innovators employed in an established industrial company to write high quality business plans. The business mentality is often foreign territory to them, yet they have to write the business plans themselves. The coaches can be external trainers or internal experts from the company's new business development and financial modeling departments.

- Professional jury and evaluation.
  This ensures that the company's financial and human resources are only invested in high quality business plans. The professionals in question will include top managers, and, if a technology-related plan relies on expertise foreign to the company, the relevant external opinion leaders need to be involved.

- A supporting infrastructure that allows the competition to be carried out on a structured and transparent basis.
  A good supporting structure is likely to include a handbook which can be ordered, downloaded, and read online, a clearly structured competition process and rules for participation, seminars, discussion forums and broad communication of successful examples.

9.3.2
Corporate Venture Funds

Corporate venture funds are venture capital funds financed partly or entirely by industrial companies. They are ideally managed using the same financial criteria as venture capitalists would apply but, in addition, they have a strong relationship with their industrial investors. The startups funded then benefit from the investors' industry knowledge, while the industrial investors not only receive a financial return but also better information from the market to use in external and internal investment decisions. They can provide access to new technologies and allow companies to build an external network by, for example, cooperating with a number of different startups. This has given some corporations access to important deals in return for a relatively small investment.

Corporate venture capital (CVC), or corporate venturing, has helped 3M, for example, to develop a number of new products. The company has gained deep insight into new technological developments, purchased the patents on promising startup technologies, concluded distribution agreements and cross-licensing contracts, and actually acquired startup companies. For example, it purchased optical lightening film technology from TIR and the rights on an aerosol for asthma treatment from Fluid-Propulsion Technology. 3M also acquired Cardiovascular Devices, allowing it access to the company's blood gas monitor, and EoTech, which provided a new glass fiber technology.

However, the historical track record of CVC overall has been poor. A survey in the late 1970s found that only 7 percent of corporate venturing programs considered themselves to be successful. Some important chemical companies were numbered among the failures.

After a long period of low activity, the commitment to corporate venture capital has increased sharply in recent years. The high-tech sector (including companies like Cisco Systems and Intel) has been enthusiastic about the idea, and corporate investment in US venture capital funds increased tenfold between 1995 and 1997 (Fig. 9.2). European chemical companies that are currently active in corporate venture capital include Bayer, Aventis, and BASF.

Recent deals in the field of agrochemicals – mainly made between 1997 and 1999 – have demonstrated that it is financially more effective to make a large number of early corporate venture investments in high risk plant biotechnology businesses, for example, than it is to acquire a successful startup at a later phase. The price premiums that have to be paid in the later phase are much higher than the cost of placing many bets widely at an earlier stage.

To unlock the potential in corporate venturing, companies should take the following steps:

• Set clear strategic objectives and be stringent about high financial returns.
  3M aims to get 25 percent of its revenues from new products, and its venture capital fund has historically yielded a 17 to 18 percent return.

- Make a strong and long term commitment.
  Many early CVC programs were too short. Our findings show that the present value of the cumulative returns only turns positive, on average, in the seventh year.
- Provide a complementary role to venture capital firms, co-investing with them or investing in pooled/dedicated funds.
  There are various approaches to venture capital for corporations, and each requires a different level of skill and involvement (Fig. 9.5). Chemical companies that want to build new innovative businesses using this method might want to start with a pooled or a dedicated fund investment to build capabilities and external reputation in the startup community. To make co-investments with VCs or to act as lead investors, corporations need to be experienced in deal flow generation and due diligence. These skills permit strategically tailored investments.
- Develop a VC team that is strongly linked with the rest of the corporation, so that it can effectively leverage existing corporate assets and will not be primarily driven by monetary returns.
  According to the experience of co-investing VC companies, the three critical skills for the corporate venturing team are internal networking, technical/market expertise, and external networking, although it is also important to have financial structuring and transaction capabilities as well as the ability to nurture startups by recognizing the most promising ones and, at the same time, make sure that they are canceled as soon as telling indications of failure appear.

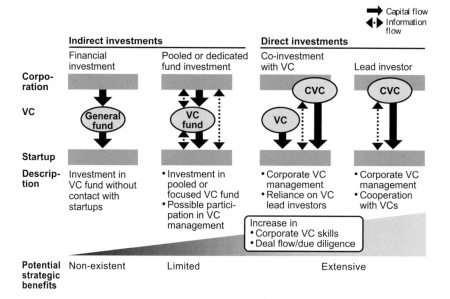

Fig. 9.5 Different investment approaches for corporations
Source: McKinsey

9.3.3
New Company Structures and Incentive Systems for Business Building

Finally, promising ideas have to be grown into real businesses. A highly effective approach here is the creation of new corporate structures and incentive systems which mimic the startup companies as closely as possible, while still providing the appropriate level of support from the parent company. Finding the right trade-off here is crucial.

On the one hand, the new business needs independence from the parent company to develop its own ideas. It must have the freedom to interact independently with the external community and have dedicated resources which are not cannibalized by the existing business.

On the other hand, it also needs strong support from the parent company, and in more ways than merely financial: access to the parent company's knowledge, internal network and business services are also required. Additionally, the parent company needs to put a controlling system in place for the new business to ensure that the investment is really giving the expected return.

A corporate structure that copies the venture capital approach seems to be the most effective way to achieve these objectives. The new business – a relatively independent entity, as described above – is controlled by the parent company through staged financing, given when pre-agreed milestones have been reached. These should include not only internal milestones defined, for example, by specific R&D achievements, but also externally driven ones reflecting important steps towards market success, based perhaps on initial results of user tests, consumer research results, competitive benchmarks, and so on. The parent company should supply services to the new business on the basis of well-defined price and resource agreements.

DuPont's Qualicon, which received several awards for innovation, is one example of a new business that was developed by this kind of approach. The idea of developing machinery and kits for automated genetics-based pathogen detection was generated by the firm's central R&D operation. It was then subsequently developed by giving it the status of an independent business unit. In another example, Dow and its partner Cargill set up an independent joint venture to push the development of biopolymers in an independent environment with clearly allocated resources and a defined roadmap and performance targets.

From the organizational point of view, startups can be given a number of different forms, and can be located inside or outside the parent organization. In each case, the formal design not only ensures the high autonomy discussed above, but also emphasizes the importance of these new businesses by creating separate units for them and, for example, having them report directly to the top level of the organization (Fig. 9.6).

Finally, a "new company" culture should be fostered in such businesses, by, for example, recruiting experienced entrepreneurial talent for the leadership team and building new incentive systems that match the rewards of a startup in the open market. Such systems should allocate a fair share of risk and return to the

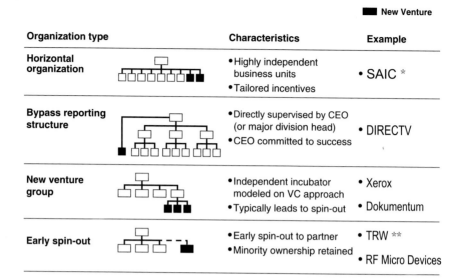

Fig. 9.6 Various forms of organization for new businesses
* Science Applications International Corporation; ** Thompson Ramo Wooldridge; Source: McKinsey

new business's employees and the parent company. The employees are normally given financial incentives consisting of stock options that guarantee attractive returns in the case of success and/or other special bonuses based on milestone achievement, and non-financial awards such as freedom to experiment and peer recognition. It is also important to specify the employees' risk in the case of failure.

On the one hand, this ensures that they are highly motivated (and are prepared to take the risk). On the other hand, it makes it easier to communicate the new structure to parent company staff remaining in the traditional career path, who will recognize that the rewards are justified. At the same time, the risk should not be so high that it is not worth taking. In companies where risk aversion is the key block to innovation, it can be limited by offering a dual career track in the first phase of business building.

As the examples above show, some chemical companies have started to drive for more innovation, although the overall mindset of the industry still seems oriented towards internal capabilities rather than the developments and opportunities in the world outside. The inventions of chemical companies fundamentally changed human life in the twentieth century. Now, these companies need to develop the ambition to participate actively in the even more fundamental innovations expected in this century.

10
Managing the Organizational Context
Karsten Hofmann and Heiner Frankemölle

In this chapter we will look at four essential organizational ingredients in the make-up of any truly excellent international chemical company. We will first consider the "hard-wiring" of the company, to examine whether certain structures work better than others in supporting a large and complex chemical company's strategic purpose. We will then focus on the role of the corporate center, and the degrees of control and independence that suit different types of businesses. Thirdly, we will turn to the "softer" side of organization and discuss the "War for Talent", to identify the ways in which chemical companies can get a more generous share of the finite pool of truly talented managers and technicians. Finally, we will examine how top-performing companies drive and motivate their organizations to make genuine performance breakthroughs. Though the make-up of the ingredients will differ in some respects for every company, common features can be identified.

10.1
Supporting Strategy by Structure

Creating the structure that best suits their strategic purpose is no easy task for the chemical conglomerates. For decades, their main challenge on this front has been to strike the right balance between leveraging size and scale in order to gain competitive advantage, whilst at the same time remaining nimble and responsive to the customers and markets that they serve.

Although these companies all face comparable challenges, they have produced a bewildering variety of organizational structures. Consider, for example, DuPont, where 24 strategic business units report to eight global vice-presidents in a structure where there is no specific regional or country representation in the main leadership group (the DuPont Business Strategy Council). Then compare that with the 17 operating divisions, 12 regional divisions, and 15 corporate and divisional functions reporting to eight board "Ressorts" at BASF. Or visualize the 15 business areas reporting to four business segments at Bayer, where the members of the board are multi-functional committee members – some of them representing both businesses and regions in parallel.

Organizational research has not yet provided a clear path resolving the dilemma of global reach and local responsiveness. The first solution proposed, the matrix structure, was enthusiastically embraced by chemical companies in the 1970s, with varying and often frustrating results. Some companies still believe in the matrix for geographically dispersed multinationals, because of its ability to satisfy the demands of local markets while maintaining a global vision. Others insist – in line with the seminal book by Christopher Bartlett and Sumantra Ghoshal, *Managing Across Borders* – that the matrix organization leads to management processes which are "slow, cumbersome and costly". Bartlett and Ghoshal's alternative suggestion for reconciling the ill-matched objectives of global efficiency, local responsiveness, and worldwide knowledge management is the *transnational organization,* a structure that operates as a "mentally integrated network" rather than a formal matrix.

Developments in the New Economy, where transaction costs are falling dramatically and much looser forms of organization are emerging, have spawned even more fanciful concepts – such as the *disaggregated organization* or the *virtual organization* – to solve the organizational puzzle of products, geographies, functions, and customers.

To many managers in the chemical industry, these concepts appear too far removed from the very practical challenge of organizing their businesses for optimum effectiveness and efficiency. They are looking for a structure that is both simple and robust, and that delivers their global and regional strategies.

Our search for best practice in this area does not surface "one right answer". Nevertheless, one option does come out on top: in our experience, an organization with global business divisions (i.e., product-based business units with worldwide control of all critically important business functions) as the main driving axis seems to fit the needs of the chemical industry particularly well. This has also been recognized by many major players: over the past years, decision making authority has in many instances been shifted towards such divisions.

Global business divisions do have some disadvantages, such as a loss of some synergies in, for example, financial functions or in national sales operations. In our eyes, however, they are clearly outweighed by the advantages. The value of such global business units lies in several different areas:

- The real profitability of different products becomes more transparent as cross-subsidies among product lines are exposed. This is a problem intrinsic to the highly integrated production processes of many chemical companies.
- Clear points of P&L accountability are established, thus promoting personal ownership and responsibility for results along a clearly defined range of products – a prerequisite for a "no excuses" policy in any organization.
- Opportunities for building businesses arise, and these help to attract and retain entrepreneurial talent. This is especially important in an industry where separate product lines have very different value propositions and incentives, for example, New Economy businesses such as business-to-business (B2B) marketplaces for chemicals versus Old Economy commodity-type businesses such as polysterol.

- More than 40 percent of chemical sales are outside the companies' respective home markets, and all the industry's major players have a global presence. In such a context, global business divisions allow for coordinated product strategies and customer service across the world's markets.
- Many internal service providers have to serve a number of different masters with quite diverse demands. Business divisions allow the trade-offs between these demands to be optimized, often resulting in substantial savings and higher quality service.

To set up such global divisions successfully, companies have to design a number of important elements very carefully up front. While the most critical element of all – the role of the corporate center – is dealt with separately in the next section, the following four are close to the top of the list:

The criteria for defining the scope of the business divisions. A key decision has to be made about how best to group the organization's existing businesses. Should they be aggregated according to product or market segments? An "affinity analysis" – that is, a grouping of all functions according to their ability to maximize the overall potential for synergies, customer satisfaction, and growth – is the preferred method for comparing alternative scenarios and calculating their value. In practice, however, complications may arise if a product-oriented organization turns out to be the preferred option, yet some key customers are served by several divisions. In such cases, the management of these key accounts may be coordinated by the division with the largest sales volume or profit potential, and this should complement the otherwise product-based organization.

The degree of divisional versus regional responsibility and accountability. The balance of divisional versus regional control can vary from the extreme of complete independence for the country managers to a situation where they have very limited control over their operations. Balanced scorecards (or a simple set of joint key performance indicators) can help to build a reference system in which both the divisional and the regional units are held mutually accountable for achieving their joint targets.

Where one particular national market is extremely dominant within a global setup (e.g., the US market for pharmaceuticals and agrochemicals), it may be wisest for companies to locate the divisional headquarters in that market. BASF Agro, for example, moved its global headquarters to Mount Olive, NJ following its acquisition of American Cyanamid.

The difference between developed and less developed businesses. Chemical companies may want to make an exception to the principle of global business divisions in the case of emerging markets. In the startup phase of a business, it can make a lot of sense to focus the attention and the responsibility of managers on local or regional business development. Appointing a "VP China" or a "VP Middle East" or an "Asian Pacific Council" consisting of SBU and country managers can help companies to tap into the synergies of shared infrastructures in the target markets.

The difference between commodities and specialties. In commodities, businesses reach the critical size required for an independent product division almost by definition. The strong drive of economies of scale and scope would otherwise quickly put them out of the game. With specialties, however, the situation is different. Small businesses may well meet the market conditions required for independence, yet still be of a scale to profit from more shared services.

One way to address this issue is to develop a view on the wider business arena of which the smaller specialty business forms a part. A number of smaller business units grouped together may reach sufficient size to justify a joint "corporate center" at the divisional level. This has additional advantages: for example, it counteracts management's tendency to neglect smaller specialty businesses somewhat, especially in hybrid companies with a mix of commodities and specialties. Such a broad perspective might also initiate strategic thoughts about strengthening the business portfolio – for example, by acquisitions within the defined arena.

Some recent discussion centers on the impact the New Economy will have on the structure of chemical companies. We believe that the Internet will contribute to the disaggregation of the value chain touched on above, opening up opportunities for chemical companies to take on board partners for those parts of the value chain that can best be served by others. However, given the size and global reach of chemical businesses, we also think that global business divisions will remain at the core of these network-type configurations for some time to come.

10.2
The Corporate Center – Lean but not Anorectic

In multibusiness companies, the invariably skeptical business units and the company's concerned shareholders often suspect corporate headquarters of being a worthless overhead, adding no value. The corporate center finds itself having to justify its very existence. To ensure its usefulness and acceptance, the corporate center should put value at the absolute heart of its design: how, in a word, can the center add incremental value above and beyond that which the businesses can create on their own?

There is a very useful framework available for classifying the potential roles open to the center, based on a matrix which plots the answers to two key questions: "What is the nature of the corporate governance model?" and "How highly integrated are the SBUs?" The resulting role models range from the hands-off financial holding company, through the strategic architect and the strategic controller, to the hands-on operator (Fig. 10.1).

In the financial holding company, the corporate center focuses on a narrow set of value-adding activities such as portfolio management and strategic planning. The SBUs are managed for the most part as fully independent entities, and the corporate staff is kept to a minimum, often fewer than 50 people. Sterling could be cited as an example of this governance philosophy.

Most of the big chemical companies have chosen a multibusiness management approach, where the primary governance role of the center is to enhance the per-

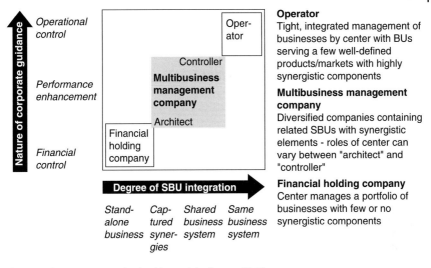

Fig. 10.1 Corporate center leadership models. Source: McKinsey

formance of the SBUs and to promote cross-SBU synergies, with varying degrees of financial and operational control.

At the other end of the spectrum, the corporate center acts as an operating company, with almost full control of operations and widely shared and centralized corporate services. In such a model, the corporate center's staff can be numbered in hundreds or even thousands, and the heads of the business units are more like line managers for their respective product groups. Bayer has traditionally been managed according to this governance model.

Obviously, the choice of an appropriate design for the corporate center – and the definition of its role – depends on a number of variables, including the strategy, the core competence of the company, and the relationship and interdependence of its business units.

We believe that three design principles hold the key to building a successful corporate center in the chemical industry:

- Small, but high caliber staff.
- SBUs run as independent entities with full P&L responsibility, each one set up as a company within the company, with full control of the business system functions and costs critical for its success.
- All non-core service functions outsourced to third parties.

Companies that have bought into the idea of a fairly small corporate center with limited day-to-day control should be careful not to become anorectic, however. With the critical size of a chemical company's market capitalization now in excess of USD 10 billion, many companies have tended to become like conglomerates of related businesses. In order to add value (and to minimize the discount that the market puts on conglomerates), these organizations' corporate centers – aside from being staffed with outstanding talent – need to be large enough to give insightful guidance to the divisions. The ideal center for a large chemical company should thus, in general, be somewhat larger than the classic "lean" center used in other industries, which sometimes has fewer than 70 staff, but – based on our research – it should not consist of more than 150 people.

The role of the corporate center should be to shape the organization, to set aspirations and new directions, to promote talent and knowledge management, and to manage external constituencies. It should be strongly focused on those key governance functions that allow it to create value for the corporation. These chiefly include corporate finance (financial control, risk management, taxation, and treasury), external affairs and investor relations, legal and corporate affairs, (group) strategy development, and corporate human resources. Optionally – depending on the importance of these functions in delivering against the overall targets and on the specificity and competence of the SBUs – the center may also include small units for R&D (mainly for new technologies and processes), IT/e-commerce, and a new venture group.

Let us take a closer look at the split of responsibilities and the range of activities normally found within the core functions of an excellent corporate center.

Corporate finance must maintain a decentralized structure, with a small central finance unit to ensure a consistent financial management system, financial optimization on the corporate level, and an unbiased view of the quality of the SBU performance parameters. The central unit should have a "dotted line" relationship with the SBUs. All generic, transaction-based functions (e.g., customer billing) should be carried out in the business units, or be outsourced. The central finance unit should focus on cross-SBU value adding services (e.g., risk management, tax optimization, growth funding) and cooperate closely with third party providers. The center should rotate personnel in and out of corporate finance, SBU finance, and operational positions to build financial expertise and strong business skills.

External affairs and investor relations should have a small group of experts at the corporate center, and the SBU external affairs personnel should have dotted line relationships with it. The corporate unit should focus on corporate identity to create a consistent and positive public image of the corporation. It should use "issue management" to shape public opinion, to pre-empt crises, and to cultivate a strong image and reputation for the company, and it needs to deal openly and honestly with crises when they occur. All external inquiries should be handled by the central unit – with the exception of public relations initiatives directly related to marketing efforts in the divisions (e.g., the launch of a new product).

Legal and corporate affairs must provide the legal advice needed to support key corporate functions and decision-making (e.g., consultancy on corporate lawsuits). In our experience, the unit itself can be very small, but should encompass enough internal expertise to segment and assign work to the most appropriate external law firms. Individuals should always be clearly responsible for the SBUs, and the lawyers may even be colocated at their SBUs' headquarters.

Strategy development should focus on corporate vision, values and aspirations, and on such strategy as is different from (but which builds on) the sum of the

SBU strategies (e.g., creating an e-commerce purchasing platform across SBUs). In the same vein, it must investigate growth opportunities that are beyond the reach of the individual SBUs (e.g., data-mining or patent-licensing companies that utilize the entire portfolio of businesses). It can also add significant value by coordinating cross-SBU networks for sharing ideas about capturing of synergies (e.g., on supplier strategies), and identifying and promoting synergies by itself (e.g., through co-branding or the use of joint sales offices). Finally, corporate strategy development needs to define the role that M&A is to play in the corporate portfolio, and should assist in SBU acquisitions (e.g., through due diligence or valuation). However, it should leave most of the generation of ideas about acquisitions to the SBUs themselves.

Corporate human resources should ensure conformity in standards across the businesses – for example, in employee evaluation processes, incentive plans, benefit plans, and training policies. As in other functions, all operational tasks should be outsourced or operated as shared services at SBU level – for example, payroll, human resources information systems, a database on internal job opportunities, or the administration of benefit plans. The center should be active in developing top talent across the businesses, and it should run the "goldfish pool" for internal top talent and the company's program for recruiting experienced personnel. This responsibility also includes the early identification of skills needed by the entire organization, for example, e-commerce skills to kick-start new businesses. The center must also act as a repository of expertise on internal and external best practices in talent management.

All decisions regarding the strategy of the SBUs, for example, capacity additions, joint ventures, long term supplies, as well as product management – pricing, product mix, quality/specifications, service levels, sales channels, marketing, product development, SCM management – should be clearly within the decision making authority of the (global) SBUs. Responsibility for plant operations and scheduling may be inside or outside the SBU structure, depending on the degree of shared production facilities.

If the corporate center is to be strongly focused on essential governance functions, where do the company's service functions fit in? In principle, there are four alternatives:

- Decentralization into the SBUs (with some form of coordination provided by the center);
- Allocation to a single SBU (on the "major user" principle);
- A central services division;
- Spin-off into a separate entity, an industrial "park" for infrastructure services.

Chemical companies have taken a number of different routes. Hüls and Hoechst, for example, before their respective mergers with Degussa and Rhône-Poulenc, integrated a great many corporate service functions into their divisions and spun off all infrastructure functions into a separate site management company ("industrial park"). Monsanto has moved its central services into a central business division called Monsanto Business Systems, while Bayer favors a strong corporate center with five corporate divisions and seven central service divisions.

While the right model depends on the individual organization's specific circumstances and strategic focus – for example, it will be more difficult to dismantle the corporate service functions in a "Verbundkonzept" such as BASF has established at all its key sites – we strongly favor a high degree of transparency concerning the level of demand and the cost involved in each service. We therefore do not advocate a central service division, since "true" market prices are hardly ever obtainable centrally. From the cases we studied, rigorous decentralization of services into the SBUs combined with total separation of the infrastructure functions (e.g., site management, safety and health incidents, operations and maintenance), seems to provide the most transparent and market-based solution. If the divisions are then given the freedom to shop around for more cost-efficient or higher quality services, this puts the right pressure on the remaining internal service providers, and, in our experience, it does lead to substantial performance improvements within a relatively short time frame.

Optimizing the structure of the organization, however, is only part of the challenge of creating a truly excellent company. Some companies excel despite deficient structures, and others fail to flourish although their organizational design has been meticulously thought through. In other words, excellent companies depend even more on the right blend of excellent and highly motivated people.

10.3
The War for Talent

Good management depends critically on a company's ability to attract, develop, and retain top talent. Indeed, superior talent will be tomorrow's prime competitive advantage. The battleground for this top talent – managerial as well as technical – has widened in scope in recent years. Today, companies are not only competing against others from within their own industry, but also with firms in different industries. The New Economy, where share options and the promise of large windfalls are the norm, has caused the war to escalate still further. Excellence in managing human resources processes is therefore becoming an increasingly significant factor for success.

In order to understand the magnitude of this War for Talent, we studied 77 large companies from a variety of industries. Nearly 400 corporate officers and 6000 senior executives were surveyed in the USA alone. More than 20 case studies were made of companies widely regarded as being rich in talent, and 5000 alumni of eight top universities participated in additional research that focused on the war for technical talent in the New Economy [1].

The main finding of the research was striking and has major implications for the chemical industry: companies are engaged in a war for executive talent that will remain a defining characteristic of the competitive landscape for years to come. Many companies are already suffering from shortages: 97 percent of the corporate officers surveyed said they perceived a strong need to strengthen their talent pool. One clear indicator of the phenomenon is the fact that executive search firms' revenues have grown twice as fast as GDP over the past five years.

For several reasons, the War for Talent can only intensify: the number of 35- to 44-year olds is going to decline significantly in many industrialized countries (by as much as 15 percent in the USA by 2015). More transparent and fluid job markets make it easier for talented people to find and name the job of their choice. These people will therefore change jobs significantly more often than they did in the past. The average executive today will work for five different companies – but in another ten years that figure will have risen to seven! Of the graduates in the survey of the last five years, 42 percent spent less than two years with their first employer. Between 1971 and 1990, the comparable figure was a mere 23 percent.

Small and young companies are increasingly attractive to top talent. 52 percent of the graduates surveyed are working in companies with fewer than 500 employees, more than double the percentage of a few years ago. Likewise, the percentage of graduates working in companies that are less than five years old has skyrocketed from nowhere to approximately 40 percent.

At the moment, the chemical industry is not faring very well in this struggle. Chemical companies are considered unattractive employers by many high potential individuals. Surveys show that chemical companies do not make it into the top 50 "ideal" employers as rated by graduate students, either in the USA or in Europe [2]. While this is partly because the entire industry is not perceived as "sexy" compared to a number of alternatives such as consulting or investment banking, graduates are also concerned about the traditionally strong hierarchy and (perceived) slow career development in the chemical industry.

Talent management therefore has to become a top corporate priority, for chemical companies even more than for others. Executive talent is, in general, an under-managed corporate asset. Our survey data show that, on average, fewer than 10 percent of executives say that their companies develop people effectively and move low performers quickly. In the chemical industry, companies manage their physical and financial assets with some sophistication, but with few exceptions they have not made people a priority in the same way.

Companies seeking to tap into superior talent as a competitive advantage must instill a talent mindset throughout the organization, starting at the top. Leaders with such a mindset spend a substantial part of their time in recruiting, training, and reviewing the performance of their executives. Companies must insist that their line managers are accountable for talent. At Monsanto, for instance, a significant proportion of a senior executive's bonus was based on his or her skill at managing people.

To attract and retain the best people, corporations must also create and perpetually refine a clear "employee value proposition" – that is, a convincing answer to the question of why any smart, energetic, and ambitious individual would want to work for that company instead of any other. Once this answer is established, attention can shift to recruiting top talent and to developing it into an even more valuable workforce.

Creating a winning value proposition means tailoring a company's "brand" in the recruitment market and making the jobs that it has to offer fit the specific people it wants to attract and to keep, all of which is easier said than done. Think, for example, of the difficulties experienced by established chemical companies in fighting against small startups for "unconventional" biotechnological talent or in persuading managerial talent not to move on to higher-paid private equity or consulting positions.

What such firms can do to help themselves, however, is to create exciting jobs for their target executive group, offering both challenges and rewards:

- Jobs that allow them to make decisions without constantly seeking approval from others;
- Jobs that provide a clear link between their daily activities and business results;
- Jobs that stretch but do not defeat;
- Great colleagues; and
- Highly competitive compensation with a considerable spread between top and average performers.

What is more, large companies do have some natural advantages – the magnitude of their impact, the large resources which enable them to take risks and to back big decisions, and the variety of experiences that they can offer. They can also help themselves by installing a small company atmosphere, for example, by creating smaller and more autonomous units, by offering startup-type environments in growth areas, and by granting substantial rewards for top performers.

When the employee value proposition has been clearly established, the focus can shift to developing a best-of-breed sourcing strategy for top talent. This entails having a very clear understanding of the kinds of people that the company really wants, the best channels and processes for finding and hiring them, and the total commitment of the entire organization to getting them.

Our research into what companies that excel in recruiting do shows some common patterns:

- Excellent recruiters put a high level of top executive resources into the recruiting process;
- They identify the sourcing pools and sourcing strategies that work best for them;
- They recruit continuously, not simply to fill openings as and when they arise;
- They do not rely exclusively on a single sourcing strategy;
- They hire from outside occasionally in order to refresh the gene pool, even if their main focus is on the internal development of talent;
- They are creative. Enron, for example, recruited hundreds of retired military officers because their geographic flexibility suited the requirements of the firm's international operations.

To develop and retain top talent, chemical companies must perfect their internal human resources practices. Our research confirms some things that many human resources professionals have been aware of for years:

1) A key to development is to give good people big jobs before they expect it. Only 10 percent of the executives surveyed strongly agreed that their companies used assignments as an effective development lever.

2) Extensive feedback and coaching is an important facet of developing people. While most companies have appropriate systems in place, execution is often not as good as it should be. Only 30 percent of executives rate their company as "excellent" or "very good" at providing informal feedback and coaching; and only 25 percent are satisfied with the way they are mentored.

3) The value of the attention of top management in the process of assigning people to jobs is underestimated. Great companies show great concern for their people. At Emerson Electric, the CEO becomes personally involved in decisions on filling the critical positions three to four levels down in over 50 divisions. Moreover, the top 1000 managers' color-coded profiles are displayed on the walls of a room at the corporate headquarters so that various staffing options can be played out while seeking optimal staff rotation.

4) Last, but not least, it is critical to move on under-performers quickly in order to avoid a vicious cycle with enormous cost. It is incredibly demoralizing to the rest of a team if poor performers are not moved out, and it makes the leader look blind and out of touch. In our survey, 63 percent of the executives at companies with average performance agreed that the least effective people tended to stay in position for years before the company took any action. We believe that taking action to deal with poor performers is the most difficult (and also the least exploited) talent-building lever for any company, and that is particularly true for big chemical companies. They have traditionally been very strong on nurturing the mentality that their people will be taken care of whatever happens.

10.4
Corporate Culture: The Key to Top Performance

Even when a company has the right strategy, the right structure, and the right people, it may still not be one of the small band of truly excellent companies – the ones that outperform the competition year in, year out. We believe that the secret of top performing companies' success, whether in chemicals or other industries, lies in management systems and processes that build up a distinctively strong corporate "performance ethic".

To find out more about the ways in which these companies build their performance ethic, we interviewed senior executives at a number of "Old" and "New" Economy companies, all of whom had chalked up many years of outstanding performance in their industries. We also interviewed 325 executives who had "grown up" in high-performing companies and then left to become CEOs or COOs of other organizations [3]. They came from 50 companies – including some from the chemical industry – that had been praised for fostering talent and that were high performers (with a 24 percent average total return to shareholders from 1986 to

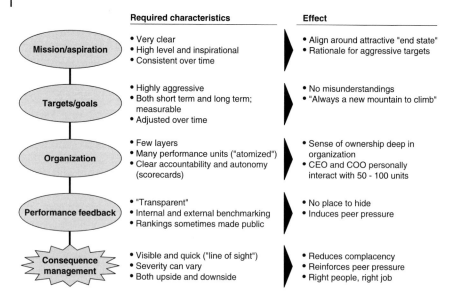

	Required characteristics	Effect
Mission/aspiration	• Very clear • High level and inspirational • Consistent over time	• Align around attractive "end state" • Rationale for aggressive targets
Targets/goals	• Highly aggressive • Both short term and long term; measurable • Adjusted over time	• No misunderstandings • "Always a new mountain to climb"
Organization	• Few layers • Many performance units ("atomized") • Clear accountability and autonomy (scorecards)	• Sense of ownership deep in organization • CEO and COO personally interact with 50 - 100 units
Performance feedback	• "Transparent" • Internal and external benchmarking • Rankings sometimes made public	• No place to hide • Induces peer pressure
Consequence management	• Visible and quick ("line of sight") • Severity can vary • Both upside and downside	• Reduces complacency • Reinforces peer pressure • Right people, right job

Fig. 10.2 Be world-class at five "must do's"
Source: McKinsey

1996 versus 14 percent for the Fortune 500 and 13 percent for the Dow Jones industrial average). In addition, we talked to a host of managing partners from private equity companies with a reputation for turning around the performance of acquired companies fast, and to a selection of the most renowned academics in the field.

We found that there are, essentially, five things that an aspirant to top performance must do (Fig. 10.2). While most chemical companies do all of them to some degree, the top performers execute the five elements at truly world-class levels:

1. Mission and aspirations. These need to be very clear, very inspirational, and very focused around an attractive ultimate goal or "end state" – for example, becoming an industry leader. Ideally, they should also provide the rationale for "stretch targets" (see the second "must do"). Most chemical companies have recognized this and have invested considerable time and money in coining their missions and their aspiration statements. However, only a few of these are both compelling and provide a clear view of the future that can be used to derive a set of meaningful targets.

2. Targets and goals. Many companies set future targets on the basis of past history plus some, with the "some" taking into account the general growth trend plus a dash of aspiration. One of the most important things that high-performing companies do differently is to set very aggressive short and long term targets for themselves that link to aspirations and can easily be adjusted upward over time in order to inject extra motivation.

In addition to their baseline targets, which are typically tied to annual budgets, top performers will also set "stretch targets" that move the organization out of its comfort zone. While everybody is expected to meet the baseline goals, those who hit the stretch targets reap big rewards. Some chemical companies go even further than that and establish best-in-class performance targets which are typically not dependent on timing. Zero defects, or the achievement of technical limits and/or core costs, are examples of these.

There is ample evidence to counter cynics who argue that it is not really possible to transcend normal performance limits.

- After discontinuities (such as LBOs, spin-offs, acquisitions, etc.) companies sometimes reach dramatically higher performance levels within a very short time frame;
- Almost every company has experienced project teams that have far outperformed others with comparable starting conditions – particularly in knowledge-intensive fields;
- There are (and always have been) extremely wide ranges in individual performance levels.

Such performance disparities are too widespread to be denied. Yet many companies still seem to consider them to be exceptions rather than using them to set their own overall corporate goals.

3. *Organization.* Top-performing chemical companies create many SBUs with P&L responsibility, or at least with detailed scorecards tied to their performance. The goal is to have absolute clarity of accountability and full visibility of performance results. The CEO and board members interact frequently with each unit, using hard-wired systems to track performance, and spending a significant amount of their time on this activity.

4. *Performance feedback.* High-performing companies like DuPont or Eastman Chemical are also excellent at providing feedback. Companies that excel in managing performance measure units against external benchmarks (the capital markets, their industry peers, the best-in-class performers from other industries, etc.) or internal benchmarks. They then rank them and many make the results public. Most of them prefer to motivate through positive reinforcement.

5. *Consequence management.* Top companies move more swiftly and aggressively in handling under-performing staff. They are more prone to use peer pressure to force the under-performers out through self-selection. They are less willing to invest in counseling. An inability to provide a significant spread between high and low performers and to react decisively on performance problems poses a significant risk to company morale.

While each of the five "must do's" needs to be carried out well, managers can make choices on two critical dimensions: first, how they coordinate and control performance, and second, how they motivate the organization to perform. Within each of these dimensions, management has to decide which of a set of levers should be pulled at distinctive levels rather than "just" very well.

In the *coordination and control* dimension, these levers are people development, financial planning and control, and operational planning and control. Most chemical companies have established financial and operational planning processes that appear rigorous and well thought through. The difference at the top-performing companies lies basically in the way that the planning and review sessions are held. They are not polite occasions where presentations are rubber-stamped. They are, rather, very intricate discussions which create an atmosphere of constructive tension and which lead to a true dialogue about performance. On the people development side, not many chemical companies have a rigorous, comprehensive, world-class performance management process in place such as the one implemented at PepsiCo/Frito-Lay.

Our research into the *motivational* dimension shows that the immediate context of the individual's work group has the greatest influence on performance. Leaders should therefore focus much of their effort on influencing this context rather than on shaping the "corporate culture" as a whole.

This does not mean, of course, that "high-level" actions – restructuring incentives, redesigning evaluation processes, and so on – can be neglected. It remains critical to get them right. A locked-in spread between high and low performers can demotivate even the most talented, top-performing team. The way in which these formal actions are interpreted, however, and the impact that they have on behavior, depend crucially upon the values, mental models, interaction patterns, use of language, and so on that prevail within each work group in the company.

Is it possible to come to grips with something so intangible, so elusive, and so complex as a group's context? Our research suggests that it is, and that the group's context can be influenced on the basis of an understanding of four main forces:

- The quality of direction: do the members of the group have high levels of understanding, conviction, and enthusiasm with regard to the target and the plans of action? Do they perceive a match between their own personal goals and the goals of the company?
- The quality of interaction: does the pattern of collaboration within the group bring about higher levels of trust, energy, and mutual accountability?
- The quality of renewal: is there an underlying dynamic to the group that enables it to develop in terms of its knowledge, performance, and innovation?
- The quality of rewards: is compensation attractive and tightly linked to individual performance? Are there also strong non-financial benefits in place?

The crucial point that has to be understood here is that incremental shifts in these forces do not work. There are thresholds in each of them that can only be crossed by great leaps. For a group to make such a leap and transcend its self-imposed performance limits often requires the presence of a "disturber", someone who shakes the group out of its routine. The disturber brings a profoundly realistic (and often painful) insight into the group's current state, a recognition of how open and inspiring a different future could be, and a sense of the thrill that a surge in performance would bring to the members of the group.

To stimulate such visions requires a deep understanding of what group members actually think and feel. Highly personal interviews can be used to find out, and these interviews may need to run for hours in order to expose the deep structures of people's beliefs. Detailed performance metric surveys can be employed to bring to the surface a comprehensive picture of the group's thinking. In intensive dialogue workshops digital voting systems can be used to achieve full transparency of the individuals' beliefs and to stimulate intense discussion within the group. All this has to be closely linked to the central business challenges, because talking about a group's performance separately from the overall business performance makes no sense whatsoever.

We advise chemical companies to take a sequential approach to implementing a strong performance ethic. First, gather the facts about the current performance ethic and identify any gaps compared to the five "must do's" and the levers. Then fill in any gaps in the "must do's", and work to achieve at least average performance with all of the levers. It is much more important here to do a few things really well than to try to "boil the performance ocean".

Once a chemical company is making good progress in filling the gaps, it should choose at least one lever and design a plan to make it truly distinctive. The choice of lever will depend on the senior team's preferences, existing strengths, and the company's financial and strategic position. For instance, virtually any company faced with serious financial trouble will pull the operational, planning/control, and incentive levers, because they are fast-acting and easy to monitor.

Embarking on such a journey requires a tremendous effort. Three quarters of the managers in our survey agreed that it was quite difficult to improve the performance ethic of an organization. But then, who said becoming a truly excellent company would be easy?

References

1 CHAMBERS, E.G., FOULON, M., HAND-
   FIELD-JONES, H., HANKIN, S. M., MI-
   CHAELS III, E.G.: The War for Talent,
   The McKinsey Quarterly (1998) 3:44–57
2 The Universum graduate survey 1999 –
   Pan-European Business Edition; US-Edition

3 STRICKLAND, W., ELDER, R.: Performance
   Ethic – Out-Executing the Competition,
   McKinsey & Company, Inc., 2000

11
Creating an Entrepreneurial Procurement Organization
Khosro Ezaz-Nikpay, Ulrich Horsmann, and Helge Jordan

World-class purchasing and supply management (PSM) is still one of the most important untapped potentials for many companies. Its impact goes beyond simple – though important – cost cutting, as productivity improvements in this area will also impact operational efficiency, innovation rate, and growth. In this chapter, we will outline the major impact that world class PSM can have when linked to an organization's key strategic challenges, outline ways to plan a journey that will create a deeply entrepreneurial purchasing organization, and make a few suggestions on how to get started.

11.1
A Bold Move in Purchasing can Create Huge Value

Over the past decade, a number of industries have systematically set out to improve their PSM. For many of them, the focus on PSM has been stimulated by the successes that the automotive industry had previously achieved by optimizing procurement in the 1980s, and by the ripple effect of these successes on upstream industries like systems and components manufacturing, metals, and chemicals.

A number of companies in the chemical industry were among those which discovered the enormous value that could be created by improving their PSM capabilities, a value typically in the same order of magnitude as a major new product or business. They also discovered, however, that they needed to be continually innovative in their PSM if that value was to grow year after year. New initiatives in the area have spread so fast that techniques like cross-functional teams – a novelty only a short while ago – have now become almost like a commodity, and are no longer sufficient on their own to enable a firm to distinguish itself from the competition.

Developments in information technology have expanded the range of opportunities for companies to derive value from procurement. Electronic-enabled PSM has accelerated this trend, and it is unthinkable today to embark on a program designed to make a significant improvement in a company's PSM without taking new technology and e-business building opportunities on board. Even the big "traditional" players in the chemical industry are now moving fast to participate in the New Economy by joining or establishing e-purchasing platforms like Chem-Connect, fobchemicals, CheMatch, and so on, thereby dramatically changing the competitive landscape. However, it is absolutely crucial to make sure that excellent internal capabilities are in place first. Without a solid base of this nature, many companies investing in major platforms will fail to capitalize on the hoped for opportunities.

11.2
PSM Comes into its own when Linked with Key Strategic Challenges

When companies launch a new product, increase their market presence, or strive to improve operational productivity, competitors are often aware of these activities and follow suit. Not only do such moves typically require significant investment, but their benefits tend to be rapidly eroded by the benchmarking activity of rivals and by the steps that they take to catch up and then to overtake the first movers. In contrast, creating world-class purchasing and supply management capabilities can largely go unnoticed. Typically, productivity improvements in a company's purchasing will also impact operational efficiency, innovation rate, and growth. One of the great strategic advantages of making improvements in PSM is that the improvements remain largely hidden from a firm's competitors. In that respect it is more akin to R&D than the more overt steps that have to be taken in order to make significant improvements in marketing or production.

Opportunities in this area arise on almost all occasions when there are "discontinuities" in a business, and many top-performing companies align their PSM efforts with the challenges they face. Mergers and acquisitions are the most obvious case. However, significant savings can also be made whenever a firm enters a new market or makes a radical change in its supply base, and/or when major regulatory changes occur – in particular these days through deregulation. In new markets, strong relationships with suppliers can be vital to a successful entry. Thus, a well conceived procurement capability can help companies find more complete answers to mastering their industry's challenges. Likewise, procurement's positioning should be entirely focused on current and expected challenges to the industry's status quo. Three of the more prevalent challenges in the chemical industry and the approaches by which some companies have leveraged their procurement function to turn the challenge into a key source of value creation are described below.

Mergers & Acquisitions. Ever-increasing M&A activity is shaking up the chemical industry's structure (see Chapter 14). Between 1995 and 1999, the global transaction value rose from USD 6 billion to over USD 60 billion. Early achievement of the anticipated synergy targets is vital not only to justify the merger to investors but also to create credibility and support for the deal within the merged organizations. Most mergers fail because the cultural clash is not overcome when two entities with distinct characteristics are forged into a new body.

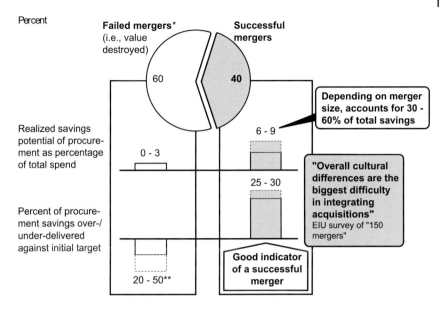

Fig. 11.1 The role of procurement in merger success
* In failed mergers, projects typically get delayed for at least a year; ** Even if a successful turnaround is achieved later, 5–10% of the initial merger value proposition is lost; Source: Economist Intelligence Unit, McKinsey analysis

PSM makes an ideal starting point for post-merger integration. Companies that have just been through a merger or acquisition (or that are planning one) should make it one of their first targets for improvement. Because the benefits generally show through clearly and quickly, it encourages other areas of the business in their subsequent attempts to draw on the synergy of the deal. This can turn PSM into a flag-bearer for the cross-cultural integration and cross-functional teamwork that is so vital to any deal's success.

A closer investigation of mergers shows that there is a close correlation between successful mergers and improvements in PSM (Fig. 11.1). Roughly two out of every five mergers in process industries are successful in the sense that they create value; the rest, the failures, actually destroy it. In successful mergers, the savings on procurement (which typically amount to anything between 6 percent and 9 percent of total spending) represent a significant chunk of the total savings that emanate from the deal. In successful mergers also, the savings on procurement typically help exceed the firms' pre-merger targets by a significant amount, a good early indicator that the merger is going to be a success.

Gains from PSM are particularly significant in the first year after a merger, the time when the new entity is most in need of signs that it is creating value. In a number of companies that we have worked with, up to 80 percent of any savings in the first post-merger year have come from PSM (Fig. 11.2). Although early savings come primarily through price arbitrage and leveraging the new firm's pur-

Percent of total potential

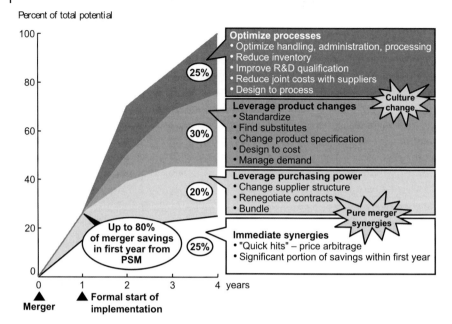

Fig. 11.2 Evolution of savings potential
Source: McKinsey

chasing power – by renegotiating contracts, for example, or by changing the structure of the firm's suppliers – the bulk of the later savings come either from leveraging changes in products or services (e.g., through standardization or by substitution) or from optimizing processes (by reducing inventory, improving administration, etc.).

Herein lies the key to why excellence in procurement is such an important indicator of a healthy merger. In order to achieve significant savings from product changes or optimized processes, in-depth cross-company and cross-functional involvement is necessary. The two "old" cultures thus start working together on creating new value and indirectly also a new culture.

Innovation. Developing a procurement capability that is intimately linked to the innovation priorities of the company can significantly boost innovation rates. Many segments of the chemical industry are experiencing increased pressure to lift their innovation rate. The focus here is not only on creating completely new products, but frequently involves innovations in product characteristics and applications development (e.g., ceramics, performance plastics) or process optimization (e.g., commodity chemicals).

Many leading firms today are also spending an ever-increasing percentage of their R&D budget on co-developments with their suppliers. The extent of this dependence on co-developments varies from industry to industry, and reflects the relative significance of material costs for a given industry. For automotive companies, for example, procurement accounts for some 80 percent of total costs. With chemicals the figure is around 60 percent, whilst for steel it lies between 40 and 50 percent.

In the automotive industry – the leader in the field – the most innovative firms today are spending up to 60 percent of their R & D budget on co-developments with their suppliers. In the steel industry the figure is considerably lower (some 10 to 20 percent), and the chemical industry lies somewhere between the two. Not surprisingly, for the more research-intensive chemical segments like pharmaceuticals and specialties, it is higher than it is for commodities. In the future this percentage is bound to rise for all industries. Developing world-class PSM thus helps firms to stay at the cutting edge of innovation.

Aiming for excellence in procurement requires firms to build strong relationships with leading-edge suppliers and to leverage their performance through alliances, outsourcing, and risk sharing. This encourages the spread of ideas from those suppliers – ideas about faster product introduction times, for example, or about new product characteristics or higher quality. We have seen a significant increase in the idea pipeline and a decrease in development times in some companies following the initiation of a comprehensive supplier development program. A reduction of around ten percent in product development times can be expected, for example, in specialty chemical and healthcare companies. The ability to share some of a project's risk with a supplier can also change the willingness of an organization to allow projects to mature (i.e., by significantly changing the option-pricing logic). Here we have a virtuous circle that is at the disposal of those companies with world-class supplier management: the higher a company's dependence on co-developments with suppliers the more important will be the ability to extract innovative ideas from these relationships, and the broader the set of joint projects with suppliers the fuller will be the idea pipeline.

Globalization/aggressive growth. This is another area currently high on the management agenda of chemical companies. In our experience, even excellent companies are frequently limited in their growth or regional expansion opportunities because their suppliers cannot grow at the same rate. As a result, quality issues, supply shortages, or delays often hamper the acquisition of new customers or entry into new markets. World-class procurement organizations typically enable such growth aspirations by developing current strategic suppliers or new sources well in advance of market entry. In addition, their companies benefit from improved global price arbitrage and qualification of new supply sources. One specialty chemicals company had mixed success in expanding into Asia, but realized after revamping its PSM capabilities the invaluable role the procurement function can play when planning significant expansion. Its ability to create market know-how rapidly (based on some current supplier relationships) and evaluate new local suppliers not only helped the company to avoid investing in an area where they would run into supply chain problems later on, but also provided two other sources unknown to top management. As a result, this company went from paying no attention to PSM to involving the head of procurement in all strategy development efforts.

11.3
Treat Purchasing Opportunities as a Business Idea

The above examples show that companies which link the development of their procurement capability to their key strategic challenges can access untapped potential while energizing their organizations. However, in order to understand the best way to mobilize the procurement function, we believe that another ingredient is essential. World-class organizations show entrepreneurial characteristics throughout their operations. Procurement is no exception. Managers in such organizations treat purchasing opportunities as a business idea, and they treat the procurement organization itself as an entrepreneurial business unit. Such an organization thus has the necessary home-grown drive to proactively propose areas of possible value creation, and will strive to link with the business challenges in order to position itself better within the fabric of the company and toward all internal customers. In terms of possible cost savings alone, this amounts to the shareholder value a new product or business would add to your company. An estimate of the bottom-line value creation potential for a typical chemical company is in the order of more than doubling the profitability or return on capital employed (ROCE) (Fig. 11.3).

The creation of an entrepreneurial procurement function calls for three elements: grabbing the entire company's attention with an attractive entrepreneurial value proposition, capturing the opportunities through a comprehensive program, and building the necessary long term capabilities (Fig. 11.4). This approach has

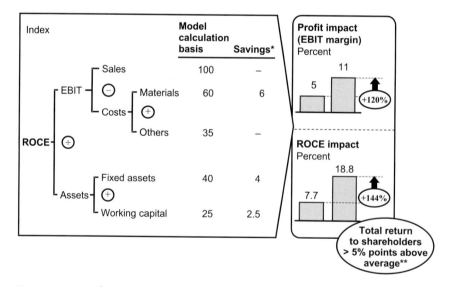

Fig. 11.3 Impact of procurement on company performance
* Assuming 10% savings on costs/assets in model calculation; ** Evaluated over five years (1994–1999); Source: McKinsey

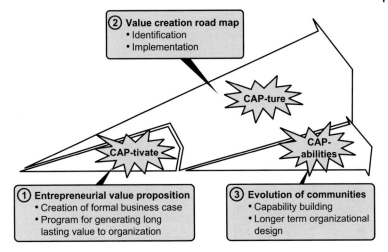

Fig. 11.4 Catalysts for Advanced Procurement (CAP) program approach
Source: McKinsey

been tested a number of times with great success, especially where the company already had an advanced procurement capability (e.g., familiarity with the total cost of ownership concept (TCO), cross-functional teams).

For example, the purchasing team at one recently-merged supplier of inorganic specialties and commodities set an improvement target six times higher than the initial synergy aspirations (initial targets were already implemented six months after the announcement of the merger). In addition, the procurement function was almost invisible to the board of management. Nine months later, virtually no board meeting passed without discussions on PSM-related topics (e.g., strategies for regional expansion, insourcing opportunities). The purchasing organization of another chemical company has developed such advanced capabilities and tools that it has formed a profit center, catering to companies outside its own group. Overall, our experience shows that if a strong entrepreneurial spirit is created within procurement the company will not only achieve significantly higher savings, but also obtain a new source of innovation and business building.

11.3.1
Captivate the Company's Attention: Create an Entrepreneurial Value Proposition

Procurement managers can captivate their company's attention by defining purchasing decisions as an entrepreneurial value proposition. To this end, they need to specify and formulate the value contribution made by purchasing to the company's overall business challenges. They then need to quantify the value added by various improvements in procurement, and to allocate that value to each function or business unit (Fig. 11.5). "Value" in this sense means the optimization of cost and quality over the lifetime of a purchase, getting more for the same price (addi-

Value proposition

- Assess value generation potential (in addition to price) and barriers

- Specify and formulate value contribution of purchasing to company's challenges

Value delivery (+)

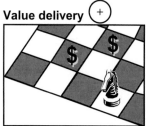

- Savings, value generation and targets per business unit/function and implementation program (budgets)

- Establish customer interface

- Valuemap key suppliers and define development roadmap

Fig. 11.5 Entrepreneurial value proposition and business plan
Source: McKinsey

tional service, conditions, etc.), or radically changing the demand (i.e., waste reduction, debottlenecking, supply chain streamlining, disintermediation).

To capture the business opportunities, the procurement team then needs to build a holistic business plan. The business plan mentality is critical, since it will provide the organization with a clear proposal and basis for the decision to allocate necessary resources to the planned efforts. Organizations that have pursued such an approach have created the prerequisites for performance transparency and service-oriented behavior. Possible outcomes depending on the needs of the company and the capabilities developed within procurement can range from a strong procurement network with its own profit and loss statement and balanced scorecard to the creation of a market-oriented profit center unit. The business plan should be focused primarily on value creation rather than cost savings, and should consist of two distinct parts:

- *The team's own development.* This involves defining the value proposition and the resources needed to make it happen (i.e., identifying the project management team and the cross-functional links that need to be built into such a team). It is important to assess the value generation potential beyond the pure "price game" and understand barriers inhibiting implementation (e.g., organizational, technological, legal/regulatory, timing barriers etc.). The team must then decide on an interim organizational and network design and draft a communications strategy for different business divisions and functions and for the firm's suppliers. Defining these interfaces and feedback mechanisms early on can save much time later during the implementation phase. The team has to set targets and estimate the resources needed to reach those targets. It should break this down into quarterly targets for cost savings and quality and efficiency improvements. Last but not least, the team should establish a learning path. Key elements of such a path are the development of a comprehensive range of products and services, the definition of capability platforms, and a knowledge development agenda.

- *The development of suppliers.* Here, the PSM team needs to establish a value-oriented framework for managing suppliers. This involves designing the right relationship with the supplier (i.e., arms-length, privileged, co-dependent) and defining what is expected of the relationship (i.e., execution, structural benefits, insight). The team should, for example, quantify the strategic value of key suppliers and draw up a supplier value map where various parameters (quality, on-time delivery, availability, warranty, etc.) are evaluated and weighted (Fig. 11.6). It should always be on the lookout for win-win ideas which will generate additional value with suppliers and should try to include sticks as well as carrots in the framework. For example, organizations should aim to build in price adjustments for those suppliers who fail to deliver value.

Such a business plan will allow more sophisticated target setting, a holistic internal customer orientation, and broader inclusion of suppliers in creating value for the company.

Parameter	Quantification	Relative importance	Supplier performance*					
			A	B	C	D
• Quality						
– Stability	• Cost of complaints	30 x	10	5	8	8
– Recyclability/ reuse	• Disposal costed • New material vs. cost of reuse	20 x	8	6	9	10
– Print/ optical quality	• Own cost vs. Outsourcing	10 x	7	7	6	8
• On-time delivery	• Qualitative evaluation	5 x	8	3	2	9
• Availability	• Cost of rush orders	10 x	7	2	3	7
• Warranty	• Cost of complaints	10 x	10	8	8	8
• Weight	• Additional freight costs	15 x	10	9	7	9
		100	90	60	70	85
• Price (USD/unit)			8	9	6	7

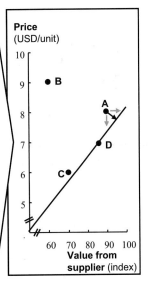

Fig. 11.6 Example of supplier value map
* On a scale from 1 (very poor) to 10 (excellent); Source: McKinsey

11.3.2
Capture the Opportunities: Embark on a Comprehensive Program

The main goal in this stage is to create a step change in the overall performance of the procurement network. Although familiarity with the TCO methodology across the business (not only within the procurement organization) will allow the launch of a more aggressive program, the key to success lies in how cross-functional teams are launched and managed. Three elements are important to make the teams work and thus achieve a successful step change: top management involvement, quality assurance, and implementation tracking.

Teams need to feel the support and the aspirations of the top management of the business. If the head of manufacturing takes five minutes out of a busy schedule to visit a team working session, for example, it might make a difference of several million dollars. Ensuring the highest quality teamwork will also be essential for a sustainable program. The program managers should continually challenge the teams (e.g., internal benchmarks, joint problem-solving, monthly reviews) to ensure that they generate a high level of fact-based aggressive decisions. Cross-functional teams that are launched in this kind of environment tend to generate many more ideas that challenge the status quo. They will also proactively apply and generate a comprehensive set of tools that are continuously refined to provide state-of-the art problem solving (e.g., Linear Performance Pricing, scorecards, risk management tools).

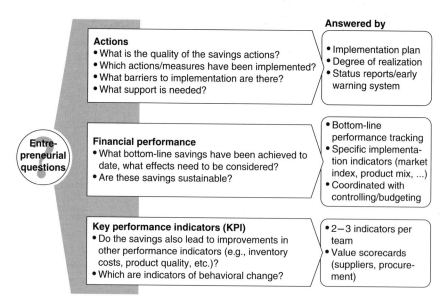

Fig. 11.7 Comprehensive tracking system – key questions and tools
Source: McKinsey

Finally, a comprehensive tracking system should be installed to measure the quality of implementation actions through milestone tracking, status of realization (i.e., started, implementation completed, money in the bank, budgeted); bottom-line performance tracking through a link to budgets and adjustments for effects (i.e., market index, volume changes, mix changes); follow-up on key performance indicators that will show fundamental changes (i.e., number of suppliers, scorecard performance, quality indicators, waste reduction) (Fig. 11.7).

In the long run, this form of tracking will also create a body of knowledge – for example, producing supplier evaluations, product performance parameters, and codified market and supplier knowledge – indispensable to the procurement organization. Although many procurement organizations are in the process of modernizing their ERP systems, data warehousing, and e-procurement environments, it is best not to wait for the optimal systems solution before implementing such a tracking system. It should be done as early as possible, and the necessary interfaces to the systems environment created later.

11.3.3
Build Capabilities: Evolve a Dynamic Organization and Design Communities of Practice

Organizations also need to think carefully about what they can do to create a dynamic PSM operation to sustain strong development beyond an initial step change improvement in their ability to create value.

In the first instance, they need to take a long term view when designing the structure of any future procurement network. The design of any network has two elements: a visible, formal structure and an enabling organization. The formal organization is typically dependent on the degree of commonality in spend behavior, supply markets, and suppliers between functions and regions. Other formal organizational elements should be designed with a view to maximizing interaction. For example, holding a managerial position within PSM must be seen by everybody within the wider organization as a natural prerequisite to further advancement. High-flyers must be persuaded to join the purchasing operation at some stage in their careers to help create the entrepreneurial momentum needed to make things happen.

In addition to the formal organization, an enabling organization should be designed with a view to leveraging know-how synergies across the network of external and internal participants. For example, individual sites may have very specific requirements for a technical item such as a heat exchanger and need specialized suppliers, thus precluding any bundling or standardization. However, sharing the value map (e.g., the performance versus price paid for certain products) across these sites will permit a valuable comparison and better supplier development. In addition, familiarity with the full range of prices paid across the network will allow target prices to be set for future purchases, and ideas to be shared on design elements. The enabling organization is highly dependent on a knowledge infrastructure (e.g., expert yellow pages, data warehouse, intranet, debriefs, and formal

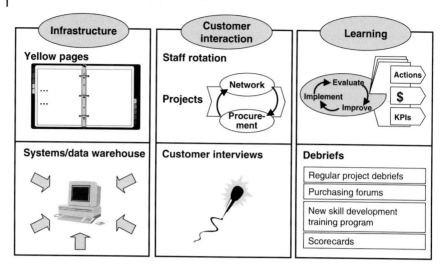

Fig. 11.8 Learning cycle – continuous improvement
Source: McKinsey

customer feedback) that will, in turn, significantly increase the development rate of the PSM organization (Fig. 11.8).

It is increasingly important, too, for firms to build the right electronic platform for PSM. E-PSM tools provide essential support for core PSM processes. Such tools might include PSM extranets with key suppliers, catalogs, on-line idea auctions, or the use of aggregators. However, it is important to remember that the main value is created internally through the better availability and exchange of information, improved ability to coordinate (e.g., to standardize or bundle purchases), or reduced maverick buying. Thus, an aggressive and entrepreneurial procurement function is a prerequisite for leveraging new tools to the best effect.

PSM organizations should set out to define those key procurement capabilities that represent a significant competitive advantage and need to evolve with changes in their companies' business environment. In the past, capabilities within a procurement function were typically single-expert based, functionally oriented and very dependent on the individual person's ambitions and opportunities. Creating entrepreneurial communities of practice around key capabilities facilitates much more rapid development of these capabilities, since groups of people with a common purpose will learn faster than individuals. These communities of practice should have a clear value proposition, and need to evolve with changes in their business environment. Examples are traders (e.g., Shell Chemicals' risk management), consortium builders (e.g., hpi, GE-TPN), innovation scouts (e.g., 3M, Toyota), or the project and knowledge coordinators that are so important for a successful procurement program.

11.4
How to Unleash the Power of Purchasing: to Boldly Go....

This is all very well in theory. But what should a purchasing manager do on Monday morning to get the whole thing moving (Fig. 11.9)?

We suggest that his or her first step should be to gather representatives of the key stakeholders and the purchasing managers in one room and hold a brainstorming session. This session should be focused on identifying opportunities for generating value from purchasing for key processes within the organization.

During the session, create "what-if" scenarios for various cash-generation opportunities. For example, consider the future based on the firm's current projected annual savings from purchasing. Then consider what would occur if you doubled the current savings. Next go even wilder and consider what might happen if you multiplied those savings five-fold. Compare this opportunity with other value generation options (Fig. 11.10).

Also, try to evaluate which of the current key purchasing capabilities are needed in the future. Then think about what new capabilities/platforms will have to be built, and about how those capabilities can make best use of developments in information technology. Finally, discuss how best to mobilize PSM within your company (on a pilot or full scale basis), and agree on the time-frame for creating a value proposition and resulting business plan and on how to get top management committed to kick-starting the effort. The new program will then be ready to go into development.

1 Gather key stakeholders and purchasing management in one room

2 Brainstorm value generation opportunities from purchasing for key processes within your organization

3 Create what-if scenarios for various cash-generation opportunities
 • current savings p.a. from purchasing
 • 2x the current impact
 • 5x the current impact

4 Evaluate which of the current key purchasing capabilities are needed in the future

5 Determine what new capabilities/platforms must be developed

6 Discuss how best to mobilize PSM within the company (pilot or full scale)

7 Discuss time-frame for value proposition/business case phase (reporting to/involvement of top management)

Use of consensor technique to surface all issues

Fig. 11.9 Seven steps for Monday morning
Source: McKinsey

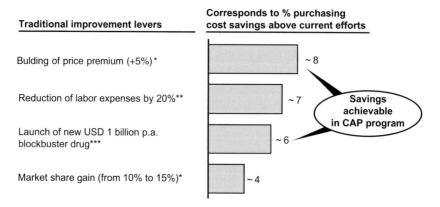

Fig. 11.10 Typical value from world-class procurement
* USD 1 billion sales, 5% profitability, total procurement spend 60% of sales; ** labor costs 20% of total costs; *** profit margin of new drug 40%; USD 15 billion sales volume, 25% profit; Source: McKinsey analysis

Creating truly entrepreneurial procurement with the high aspirations proposed is not for every company. A more traditional procurement optimization effort will still create value, but will not create a competitive advantage. However, companies wishing to turn their procurement function into a new source of strategic freedom, a business idea, and a crystallization point for entrepreneurial fervor will find it well worth embarking on this rewarding journey.

12
Achieving Excellence in Production
Markus Aschauer, Christophe Bédier, Robert Berendes, Jens Cuntze, Heiner Frankemölle, and Thomas Röthel

Efficient management of plant operations has been a key task since the earliest days of the chemical industry. However, the commoditization of an increasing number of the industry's products in the past decades and the resulting relentless price/cost pressure has made it a matter of urgency for firms to accelerate the process of improving their production and maintenance operations.

Improving production, as the biggest cost factor in a company's total expenditure, is the most effective lever for achieving operational excellence in the chemical industry. These days operational excellence can be a powerful strategic asset, since it is not easy for competitors to copy. As such, it should have a high priority on the agenda of every firm's top management. Our empirical databases on operational improvement programs indicate that excellence in manufacturing accounts for potential improvements of about nine to ten percent on a firm's return on sales, more than half of the average total potential for operational improvement of 16 percent.

Nevertheless, top managers often tend to give operational excellence a low priority. This has led to a situation where the average level of performance improvement over the last 15 years has been, on average, only sufficient to compensate for the industry's price/cost squeeze.

The levers applied to achieve operational excellence will vary among industry segments and depending on a company's mentality and position in the product and plant life cycle. The levers will range from simple cost-cutting exercises to the implementation of full-scale "Lean Manufacturing", where product and information flows are optimized. This approach alone can account for a very large share of the total savings mentioned above.

In addition to finding the right levers, companies have to undertake three major initiatives to establish such cost optimization programs successfully. They need to set themselves high aspirations geared to capital market expectations; they need to give performance a rapid boost through a significant step change; and they need to set up a program of continuous improvement to embed the performance-building mentality in their organizations on a permanent basis.

12.1

Levers for Optimizing Production Costs

In the past two decades, chemical companies in all segments of the industry have launched significant efforts to improve their production functions. The main target areas have been labor and asset productivity, and improvements here have been achieved by removing bottlenecks, expanding capacity, and building up process and product quality.

Overall performance has undoubtedly seen major improvements. Nevertheless, productivity has increased less in the chemical industry than in others: it seems, therefore, that substantial opportunities must still exist for most companies. In Europe and the USA, for instance, the annual improvements in chemicals ranged from 2.9 to 3.4 percent, whereas the figures for the steel industry ranged from 3 to 4 percent, for aluminum from 4 to 5 percent, and for automotive companies from 5 to 7 percent.

What levers, then, are available to reduce costs, and what determines which ones a given company will use? Opportunities range from functional performance improvements (e.g., in the areas of energy efficiency, personnel productivity, or preventive maintenance) to the more recently developed and even more important approach of optimizing product and information flows.

Priorities will, of course, depend on a company's sector and current cost position. Its performance culture and plant management history, as well as the stage in the product life cycle, will also play a role. At a company with a strong track record in cost saving – Huntsman, for example – even the production plants for the less cost-critical specialties will typically show a high level of energy and maintenance efficiency. On the other hand, a more technically focused plant for the production of commodities will typically be characterized by high asset productivity but only average personnel productivity, and will therefore be likely to contain major improvement potential in the latter area.

Superior cost management has been a key success factor in the more commoditized chemical segments such as polyolefines and basic chemicals (e.g., methanol) for some time now, and this shows through in strong productivity improvements.

Successful players in specialties have traditionally been less dependent on pure cost-reduction efforts, and these companies have consequently made less of an effort to improve production. In the United States, labor productivity increased by about 5 percent per annum for plastics, as opposed to around 2 to 2.5 percent per annum for paints, fine chemicals, and pharmaceuticals over the last two decades.

Improvement levers change gradually over the product life cycle, and they also reflect the plants' recent histories. In the early stages of a product's life, while it still has some of the characteristics of a specialty chemical, attention will be focused on improvement levers such as yield increases that save raw material, debottlenecking (e.g., by reducing process or reaction times), or increases in personnel productivity. Once commoditization sets in, levers like improvements in energy efficiency (e.g., by using the pinch methodology) and in maintenance and repair (e.g., by moving from reactive to preventive maintenance) become more important. In parallel,

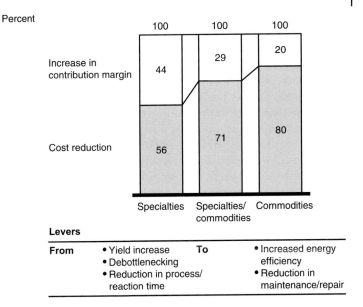

Percent

Fig. 12.1 Gradual changes in profit improvement levers
Source: McKinsey

R & D spend and overhead costs will be reduced (Fig. 12.1). Therefore, the savings potential as a share of the total operational improvement differs significantly between specialties at 56 percent and commodities at 80 percent.

As mentioned above, one other big lever is to optimize product and information flows, taking a cross-functional view of the entire supply chain from the original supplier to the end customer. This can reduce inventory levels substantially, as well as smoothing out production instability and its associated costs. It also often helps to improve levels of customer service, such as on-time delivery, and capacity utilization.

All in all, we estimate that the benefits of this approach in the chemical industry may account for up to half of the total potential for operational improvement in a firm's return on sales.

The products that stand to benefit most from such optimization are those manufactured by processes involving batch steps, and in a variety of grades that can be stored and have to be packaged. These are usually products "downstream" from basic chemicals, such as polymers, fine chemicals, pharmaceutical intermediates, performance products, and other specialties. Together, these make up more than two-thirds of total chemical output by value.

The most effective approach to optimizing product and information flows is the "Lean Manufacturing" system. Originally developed by Toyota, it is now being introduced into process industries such as chemicals after having gained wide acceptance within the automotive and other assembly industries. The overall philosophy of Lean Manufacturing is "to produce the right goods, at the right time, in the right amount, at the minimum cost". To achieve that goal, the system aims at eliminating "waste", such as excess production and wasted time, energy, raw materials, and so on by taking variations and rigidities out of the production process. Variations may include changes in throughput times or machine failures; rigidities may consist, for example, of long changeover times between successive product batches, leading to longer than necessary production runs. Both result in higher than necessary inventory levels and production costs, and often lead to sub-optimal levels of service to the customer. The typical impact of Lean Manufacturing in the chemical industry therefore consists of a significant improvement in logistics, quality, capacity, and costs at the same time (Fig. 12.2).

Most of the tools used in Lean Manufacturing can be implemented very quickly, and the results are soon evident. The following are among the most useful ones for the chemical industry:

- Pull scheduling. This ensures that production is "pulled" by demand and not just pushed, for example, by forecasts which may result in levels of stock that are either too high or too low. Pull scheduling is usually implemented through a simple but effective signal card system (the "kanban" system).
- Leveled production. This is a production planning technique which aims at aligning the product mix and lot sizes with demand in order to capture the full benefits of pull scheduling.
- Quick changeover techniques. These help to reduce lot sizes and to align them with demand without reducing capacity utilization.
- Standardization. This involves setting up standards (for example, well-defined throughput times for each process) to eliminate waste arising from variations.

Category	Achievement	Range of achievements Percent	Time to reach this performance level
• JIT/logistics	• Inventory reduction	-20 to -70	
	• On-time delivery	+20 to +25	
	• Reduced production instability	-60	
• Capacity/ productivity	• Throughput increase	+15 to +40	3 to 9 months
• Cost	• Direct cost reduction	-15	
• Quality/culture	• N.a.		

Fig. 12.2 Typical impact of lean manufacturing in the process sector – chemicals and metals based on a sample of 12 companies; Source: McKinsey

An overview of existing product and information flows can reveal whether a company's operations could benefit from the Lean Manufacturing approach. For example, if the time during which value is added is only a small fraction of the total throughput time, or if inventory levels are high, changeover times long, and the utilization rate of critical equipment low, Lean Manufacturing may provide a cure.

Given the high complexity of the operation, one of the most important factors in achieving the full impact of any improvement process is a rigorous control system that enables production managers to follow up with full and timely implementation of all the necessary measures, and to identify the bottom-line impact on the profit and loss account.

12.2
Implementing Operational Change Programs in a Chemical Plant

Operational excellence in a chemical plant is not established by a one-off effort. Typically, a number of fundamental steps (most of them sequential, although some can be carried out in parallel) require anything from three to seven years of continuous change. In our experience, three major initiatives are essential for success. First, firms need to set themselves high aspirations geared to the expectations of the capital markets, which generally implies a major change in mentality; second, they need to give performance a real boost through a significant step change within a short space of time to really energize the process; and third, they need to set up a program of continuous improvement in order to sustain better performance on a permanent basis.

12.2.1
High Aspirations

Setting high aspirations involves a complete change of mindset for most chemical companies, in an industry which has lately been content with relatively low aims. In the past, production or unit managers were often incentivized by avoidance of supply disruptions, quality issues, and safety and health incidents. Cost has played a minor role, especially in non-commodity business units. This implicit contract needs to be changed and management needs to make a conscious choice about the risk it is willing to take (with the exception of safety and health incidents). In addition, the creation of new and ambitious yardsticks provides the opportunity to communicate and establish the change throughout the organization and to win commitment, as corporate-level targets are translated into appropriate goals for the entire hierarchy down to middle management.

If chemical companies accept the capital market's expectations as their primary performance target, then attempts to achieve production excellence – and thereby influence the biggest cost element in the business system – can only benefit them. To ensure well-rounded performance improvement, the capital market perspective should also be combined with other outward-looking measures derived from both customer-related targets (e.g., significantly shortened delivery times) and expectations about suppliers (e.g., the quality of raw materials). Furthermore, an in-depth understanding of the levels of performance of the company's main competitors – exemplified in benchmarking tools such as industry cost curves – can define the lower end of aspirations and is helpful in setting stretch targets for the company's own performance.

It is vitally important to translate the various outward-looking perspectives into measurable and concrete technical targets for production managers and front-line personnel. Detailed value-driver trees are an excellent tool for establishing a direct link between financial indicators at the corporate level – such as Return on Invested Capital – and hands-on technical indicators that mean something to a production unit, such as tonnes of steam per tonne of product (Fig. 12.3).

Explaining these value-driver trees to employees helps them to understand how the individual's daily actions and performance have an immediate and direct influence on the company's (or the business unit's) profitability. Based on this transparency, specific targets for front-line and intermediate performance indicators can be agreed upon in yearly "target contracts" at all levels of the organization.

Top management has to initiate the target-setting process from the top of the value tree in a collaborative, non-authoritarian way. Targets should be set interactively in discussions between the different management levels. The process will then cascade through all organizational levels down to the front line. There, it can end either at the level of each individual production team or even of each individual employee.

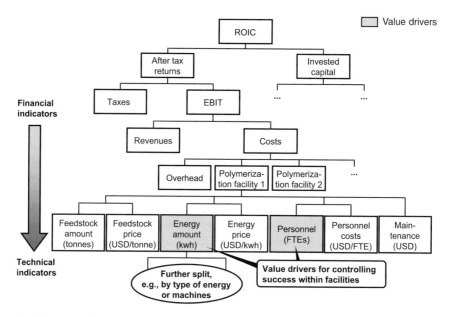

Fig. 12.3 Value-driver trees from financial to technical indicators
Source: McKinsey

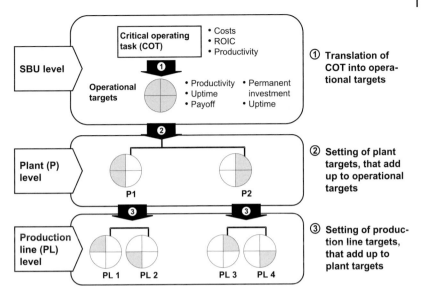

Fig. 12.4 Cascading targets from BU to plant level at DuPont
Source: McKinsey

A prime example of such an approach in the chemical industry is DuPont. The American giant has established a powerful target-setting process that starts with so-called "critical operating tasks" (COTs). These are derived from financial origins at the business unit level, and cascade down through the plant to the production line. The system creates transparency for each production line employee about his or her targets and about how the individual's performance is linked to that of the whole unit (Fig. 12.4).

12.2.2
Step Changes

Virtually every chemical production unit will improve its productivity and reduce costs to some extent over time. Typically in such cases, a number of more or less obvious improvement ideas are generated and implemented by a highly motivated but small part of the workforce. In addition, targets are mostly low, and the level of improvement that is typically reached is insufficient to compensate for the price-cost squeeze. Furthermore, the creeping improvement in performance is overlaid by segment-specific price cycles, making consistent tracking of the improvement activities almost impossible.

Breaking out of this situation requires an internal discontinuity – a real "step change" in performance, characterized by improvement targets of "inconceivable" proportions. The concept of core costs has proved to be highly successful here over the last 10 to 15 years, particularly in production environments. Core costs are the minimum production costs that would be required if all types of costs were reduced to their natural limits. This concept aims to establish a mindset which will encourage employees to accept only these natural limits as the boundaries for improvement. The natural limits, however, are a purely theoretical figure which will never be reached. To translate these into reality, companies can turn to the concept of technical limits, which is based on the idea of approaching the minimum levels of energy or raw material consumption that are equivalent to thermodynamic limits. If these are not available, a working solution can be found in the limitations of the most advanced – even experimental – technical equipment. In our experience, the core cost approach motivates employees much better than benchmarking, for example, where target-setting discussions can spark off endless arguments about why the employees' own situation is not comparable.

The difference between the present cost base and the core cost is described as the "fully compressible cost". A target – or ideal hurdle – of reducing compressible costs by about 30 to 50 percent has proved to be both stretching and also achievable in all process industries, and has been attained by many chemical companies with very different performance histories (Fig. 12.5).

A successful performance improvement initiative should take a holistic view rather than focusing on costs alone; most importantly, it should explicitly add the customer perspective. Our experience has shown that targets should be set along several different dimensions, including quality (e.g., product or process quality with a target of zero defects) and service/time (e.g., throughput times or the reliability of customer delivery) as a bare minimum. Chemical companies can reap additional benefits by adding targets for the environment and safety to the spectrum.

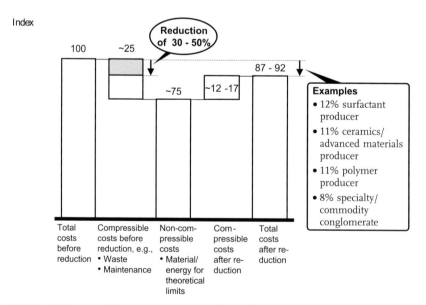

Fig. 12.5 Typical step change improvements in chemical manufacturing
Source: McKinsey

Building a step change program on targets derived from core costs inevitably comes up against typical mental barriers present in day-to-day operations – ranging from insecurity to reluctance. A winning formula has to combine stretch targets with a systematic process for generating and evaluating ideas in a relatively short time which will motivate all employees within a production unit to participate in the improvement process.

Striving for stretch targets will encourage employees participating in ideas workshops not only to challenge the normal run of operations, but also to slaughter their own favorite "sacred cows". Typically, in the chemical industry each employee comes up with an average of about five ideas. In a company, for example, with 3000 employees, that means a generous pool of around 15 000. After all, targets that shift performance nearly half way towards a technical limit cannot be achieved by slight improvements. They demand a systematic approach to generating ideas for fundamental changes in both technical processes and organizational procedures.

Finally, by applying the appropriate indicators from the value driver trees mentioned in the previous section, management can lay the groundwork for comprehensive and practical monitoring of performance improvements during the implementation phase.

12.2.3
Continuous Improvement

High aspirations and step changes in performance are both critical elements for operational excellence in production. Nevertheless, long term success eventually depends on an ongoing drive towards performance leadership. It is therefore crucial for chemical companies to institutionalize a process of continuous improvement in their production functions.

Chemical players with the very best productivity increases, firms like Nalco, National Starch, or AlliedSignal, have several elements in common. Our research shows that companies with effective continuous improvement processes share one overall characteristic: they have embedded entrepreneurship at all the various levels of their enterprises to establish a performance culture.

There are three basic elements to entrepreneurship in chemical production (Fig. 12.6). The first – establishing *entrepreneurial focus* – means that performance targets have to be set which are driven by the capital markets in a participative process starting from the top – that is, at the business unit or divisional level – and which follow through a transparent value-driver tree in order to determine targets for the front line (see Section 12.2.1). The desired end product is a balanced scorecard for each operational unit with a clear link to the overall profitability of the business unit.

Entrepreneurial opportunities require a significant part of each employee's compensation to be linked to the extent to which the organization reaches the performance targets to which it has committed itself. Making the performance-based bonus at least 20–30 percent of an employee's total compensation (and establishing

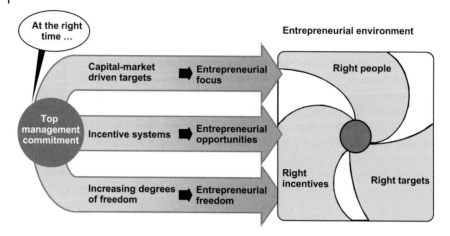

Fig. 12.6 Three basic elements of entrepreneurship
Source: McKinsey

a spread factor between low and high performers of at least two) is a highly effective way of ensuring that every employee buys into the improvement process.

Eastman Chemical has set up a powerful two-part incentive system for all its employees, not only manufacturing personnel. On the one hand, the top 600 managers have to put between 10 and 40 percent of their salary at risk. However, they can earn a bonus of up to 80 percent of their base salary depending on the economic value created, the increase in sales volume, and the improvement in labor productivity. Furthermore, they are required to own Eastman stock worth between 50 and 400 percent of their annual income.

In the second part of the scheme, the remaining 18000 or so Eastman employees put 5 percent of their income at risk and can earn bonuses of up to 30 percent of their base salary, depending on the company's return on capital. Although all employees are motivated by their company's performance, Eastman's current system still lacks a direct link between the individual's performance and his or her compensation.

Entrepreneurial freedom means that the appropriate degrees of freedom have to be put in place, particularly for middle managers but also down to the level of supervisors. Concrete examples include the total control of production costs, a leading role in investment and personnel decisions, and a strong influence on day-to-day production planning. Of course, these degrees of freedom should be combined with appropriate downside risks, including full responsibility for failures and misconduct, and a high level of responsibility for leadership and people development.

A key requirement is, as always, *top management commitment*, the critical force that constantly pushes the organization towards the other elements of an entrepreneurial mindset.

A good example of a chemical company that has achieved a very strong performance culture in its plants is Dow. Prime indicators of this are, first, the tough targets that stretch the individual businesses. These targets – with a focus on bottom-of-the-cycle forecasts and competitive cost analysis – are set by the manufacturing managers themselves. Second, Dow has achieved both massive step change improvements, such as a 30 percent reduction in processing cost during the mid 1990s, followed by continuous cost improvements significantly beyond the price/cost squeeze for the latter half of the 1990s. Third, the plant and site managers at Dow have a high degree of entrepreneurial freedom, allowing them to use all the manufacturing levers available in order to achieve their tough targets.

However, even at Dow there are skeptics to be convinced. As one manufacturing executive puts it: "Each time we get a new set of targets, I am convinced that they've gone too far. But somehow we manage to achieve them, and the business is better for it!"

The journey towards production excellence in the chemical industry takes a number of years. Nevertheless, it is a very worthwhile undertaking, as major opportunities exist in this area. Depending on a number of factors ranging from industry sector through corporate culture to the stage in a plant's or product's life cycle, levers for improvement will range from simple cost-cutting exercises to the implementation of full-scale Lean Manufacturing. Establishing and maintaining superior performance in this area depends on three main elements: setting high aspirations, making a performance-boosting step change, and then establishing a process of continuous improvement.

13

A Customer-centric Approach to Sales and Marketing

Sönke Bästlein and Jan-Philipp Pfander

Chemical companies have traditionally relied on marketing and sales to promote their product innovations, to educate customers in how to use them, and to build close relationships with these customers. This system worked well while demand grew strongly. During this period, the industry built large and complex sales and marketing organizations that were designed to serve every market segment all around the globe.

This approach, however, is now being challenged. In addition to slowing growth in most of the industry's market segments, three powerful market shifts are taking their toll: the commoditization of products, arising largely from the lack of innovation and the increase of competition from me-too players over the past decade; the increasing sophistication of customers and purchasing departments as a result of their long experience with those products; and the emergence of e-commerce which, by providing customers with a wealth of additional information, is set to redistribute power and change traditional supplier/customer relationships.

The resulting decline in competitive differentiation combined with increasing buyer power is putting pressure on marketing and sales organizations to reinvent themselves, and to demonstrate their ability to create a new kind of value for their organizations and their customers. Otherwise, they may find themselves prime targets of cost reduction programs or, at the end of the day, actually obsolete.

Is the reinvention process possible? On the basis of our experience and in-house research into best practices, we believe that it is, and that marketing and sales can still be a very strong lever for creating value in the chemical industry – if managers in these areas are able to make the change from selling products to a customer-centric approach. This approach, which focuses on selecting the most attractive customer portfolio and effectively capturing the interface to these customers, originated in the service sector in the eighties but has become increasingly popular recently in all types of industries. Its application is becoming more and more simple and widespread through the use of customer databases and the rise of the Internet, "e-enabling" the sales function. Using a customer-centric approach, companies with superior marketing and sales skills are more profitable and are growing more strongly than their peers in every chemical market. Examples of best practice include GE's engineering plastics unit, where growth is 10 percent (around 20 percent above the market average) and the Return on Sales

(ROS) 23 percent, and Loctite adhesives, where growth is 12 percent (3 times the market average) and the ROS 16 percent.

The seeds of the approach are already present in most chemical companies, especially in view of the close customer relationships the industry has relied on in the past. However, to make them grow most companies will have to change three elements in their approach and insure that concrete plans and resources are put in place for their implementation:

Customer portfolio strategy. This has to be focused on the externally and internally most attractive customers. Selecting a customer portfolio which is in tune with the company's own internal strengths reduces complexity and increases overall profitability.

Product/service mix. Building profitable long term relationships with these "attractive" customers requires a shift from selling products to offering customized solutions. Exploiting in-house knowledge to increase customer convenience has to become the key value creation lever for marketing and sales. Creating "reference" customers here makes sure that ongoing experiments can be undertaken to insure the development of product, application, and service innovations.

Management of the sales process. An absolute prerequisite for fast and profitable growth is tight control of the sales generation process. This is achieved by "programming" sales channels, controlling them on the basis of inputs so that the influencing factors can be properly understood and the right marketing and sales levers applied, and insuring that efficient service is possible by, for example, eliminating bottlenecks.

13.1
Customer Portfolio Strategy

In line with the customer-centric approach, the sales and marketing organization needs to start thinking less about what it is selling and more about who it is selling to. A renewed marketing and sales strategy should therefore begin by restructuring the customer portfolio to concentrate on those buyers with the highest profitability, growth, and sustainability. We have found that, in a customer portfolio with a typical distribution of margins and growth, a shift of only 4 percent of the volume from unattractive to the most attractive customers results in an improvement of 1.2 percentage points in ROS and of 0.8 of a percentage point in growth (Fig. 13.1).

Focusing on their growing customers allows firms to chalk up high growth even in stagnating markets. In one example, we observed that a company achieved increases of 4 percent in a market that was growing by only 0.5 percent per annum by focusing on its successful customers and increasing its share of those customers' purchases. This "leveraged" growth strategy has particular merit in markets where any gain in market share would only be possible at the risk of a competitive "over-reaction", leading to price wars.

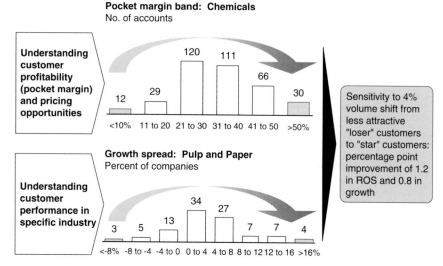

Fig. 13.1 Identifying customer opportunities for profitability/growth
Source: McKinsey

But in rejigging their portfolios, companies also have to examine their customers' attractiveness to themselves – the highest growth rate in the world will not make a customer a good prospect if he squeezes prices beyond reasonable limits or places irregular orders.

An attractiveness matrix which classifies customers by both external and internal attractiveness has proved a very effective tool here.

External attractiveness is determined on the one hand by the customer's growth rate. A further crucial element, however, is risk – how long will the customer be sustainable, or at least pursue his course without major disruptions? What lifecycle stage has his business reached? Is he – in the extreme case – vulnerable to takeover, in which case control over his choice of suppliers will be lost? The combination of high growth and low risk is, of course, ideal.

The assessment of internal attractiveness then looks at these companies from the supplier's point of view. First of all, how profitable is business with this customer? Profitability needs to take into account all the factors which drive margins for each individual customer, such as technical service at the customer site, cost through customization of products, services and all pricing components and rebates as well as the synergy effects across the portfolio – for example, the need for an outlet for any by-products, and so on. In some cases we even recommend pricing bottleneck resources, for example, sophisticated application services, at their opportunity cost.

Second, how stable is the company's relationship with this customer? This can be measured by looking into his past behavior. Is the customer reasonably loyal to a partnership approach on principle? In other words, is his buying independent

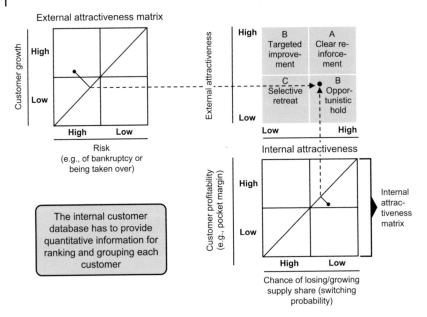

Fig. 13.2 Identifying attractive customers
Source: McKinsey

of market conditions, does he still buy from the company when he can – temporarily – get the product cheaper elsewhere, or does he behave opportunistically?

By combining the evaluation of their internal and external attractiveness, customers can be classified as "A" (clear reinforcement), "B" (targeted improvement and opportunistic hold), or "C" (selective retreat) (Fig. 13.2).

This classification leads to different strategies for the different types of customer. In general, the A customers will form the core of any customer portfolio and the focus on them needs to be particularly emphasized. B customers fall into two categories: those who are highly attractive externally but not so internally; and those who are generally only attractive internally.

Warning signals from the first of these groups need to be analyzed carefully. Too often, aggressive customers who are growing and consolidating within their industries are dismissed as unprofitable or opportunistic. However, in actual fact these customers could offer a real opportunity to capture market share by offering them something that will make it worth their while to become A customers. Understanding how to turn these customers into A customers will reveal competitive gaps within the company which have to be closed in order to be successful in the longer term. These customers should therefore be examined in detail, and strategies should be put in place to improve relations with them and, in parallel, increase their profitability.

B customers, on the other hand, who are not particularly successful in their own markets create a false sense of security due to their internal attractiveness. In the longer term, they can only lead to reduced growth and undermine the supplier's market position. Such customers should, therefore, be served opportunistically and not regarded as safe and regular buyers. With C customers, business should be actively moved elsewhere whenever more tempting options present themselves. Instead of simply not serving these customers, suppliers should outprice them. This, at the same time, sends a signal of trustworthiness to the market.

To build this classification of customer attractiveness, a comprehensive and integrated customer database is needed which allows for objective and fact-based decisions unmarred by individual preferences and prejudices. The customer ratings should always be generated across countries for the existing and potential customers of the market served by a production unit. We would emphasize that potential customers also have to be taken into account, and that some of the information used will necessarily consist of rough estimates. In view of the opportunities now offered by the Internet, we would recommend building an integrated "real-time" database as one module of the overall marketing and sales collaboration software.

13.2
Profitable Long Term Customer Relationships

Once the company knows which customers are attractive or potentially attractive, the next step is to create customized solutions for the most attractive ones, to build barriers that prevent the customer from changing to another supplier ("switching barriers").

Customers are often reluctant to see their suppliers build such switching barriers, but this move does offer them some extremely positive features. In order to achieve closer long term ties it is necessary to develop customized solutions for each and every customer. The central part of any customized solution is a distinctive value proposition that can only be offered through the delivery of significantly superior performance. To increase distinctiveness and innovativeness, a powerful approach is to form partnerships with reference customers who favor joint experiments into ways of improving their own performance. For example, Dow's Epoxy and Intermediates business follows this strategy of creating new services through getting to know customers' value chains more intimately, ideally with relatively small non-exclusive customers.

Switching barriers typically have two components:

- The one-off cost of switching suppliers – such as the reformulation of products or investment in new systems; and
- The permanent switching costs – such as the risk presented by a new supplier, or the additional transaction cost.

Both components can be used to generate barriers. However, most suppliers focus on the one-off switching cost (through, for example, joint warehousing or order fulfillment systems) because this is the traditional lever used to lock customers in.

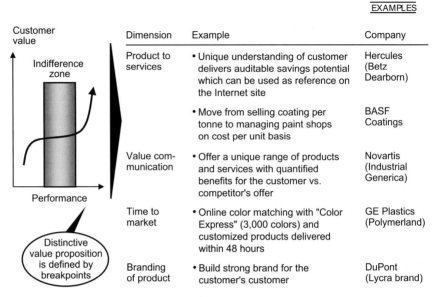

EXAMPLES

Dimension	Example	Company
Product to services	• Unique understanding of customer delivers auditable savings potential which can be used as reference on the Internet site	Hercules (Betz Dearborn)
	• Move from selling coating per tonne to managing paint shops on cost per unit basis	BASF Coatings
Value communication	• Offer a unique range of products and services with quantified benefits for the customer vs. competitor's offer	Novartis (Industrial Generica)
Time to market	• Online color matching with "Color Express" (3,000 colors) and customized products delivered within 48 hours	GE Plastics (Polymerland)
Branding of product	• Build strong brand for the customer's customer	DuPont (Lycra brand)

Fig. 13.3 Shaping breakthrough ideas
Source: McKinsey

We would argue, though, that the permanent switching cost arising from long term relationships and in-depth knowledge of the customer's organization and systems (e.g., joint working teams and process interfaces) is the more powerful barrier. Customized products that are angled towards this type of lock-in are much more effective, and also create less suspicion in the minds of customers.

Examples from different industry segments show that moving from products to services is one way to build switching barriers (Fig. 13.3). Another option is to build a strong brand in the minds of the customers' customers – as DuPont has done with Lycra, which itself has become a fashion item. But branding can only leverage – not replace – performance.

Then again, with OEM coatings BASF moved from selling product by the ton to managing paint shops on a cost per unit basis (see also Chapter 3). This offered a unique opportunity to generate significant customer value by managing the whole system, right across the traditional interfaces between the paint shop operators, the different coating suppliers, and the systems manufacturers. For the company's customers, the need to manage complexity has been significantly reduced, and systems performance significantly improved. At the same time, the supplier has had the chance to manage the complete system and to build up his understanding of what specific measures will create a lasting switching barrier.

Arguably, e-commerce will allow chemical marketers to offer information-based solutions and customization not only to a select number of established accounts but also to a much broader range of customers in profitable growing market segments at much lower cost. GE Plastics' "Color Express" service, for example, allows online color matching via the Internet for small customers, a segment of the market that is growing faster than average. By doing so, GE Plastics has reduced delivery times for these customers (who place a high value on speed) from anything up to four weeks down to 48 hours.

The focus on switching costs puts the marketing and sales organization in the right frame of mind to create additional value, which they can do by translating the switching cost into a price premium, exploiting the range of the cost of switching to the next best alternative.

It has rightly been argued that the relative amount of switching cost differs for the various segments of the chemical industry. However, it should always be possible to build some kind of switching barrier into the design of customized solutions. As our examples show, there is a very wide range available.

13.3
Absolute Control over the Sales Generation Process

To implement a customer-centric marketing and sales strategy, chemical companies have to insure that they understand and can manage the customer interface. In order to be successful here, companies need to gain absolute control over the sales generation process, from the first customer contact to post-sales follow-up and customer care, and to use a broader selection of channels, including applying e-commerce simply as a channel or as an enabling, that is, collaboration tool for the direct sales force, distributors, and call centers (see Chapter 7).

To accomplish this, the sales and marketing function has to change its processes. The present design, which relies fairly heavily on individual skills, has to shift base to a more programed and input-focused approach. In other words, instead of being content to judge sales performance by whether volume goals are met or the bottom-line profit and growth contributions are sufficient, companies have to know precisely what goes into the sales process in terms of, for example, visits, time spent with the customer organization, different skills applied and hit rates as well as finished goods inventory in hand, so that they can identify the levers which will improve their customer service – and thus volume and margins – even further. Moreover, this change has to apply not only to the management of the various channels, but also to the management of those internal functions critical to sales generation – for example, the internal application laboratories, which so often restrain growth because the most knowledgeable and effective resources are not identified as critical, and their deployment is not steered by the contribution at stake.

The need to approach a defined customer portfolio with customized solutions through coordinated multiple channels will lead to a stronger role for strategic marketing, which will be empowered and brought closer to the market through much more detailed information flows as released by advanced customer databases and collaboration tools. The future role of strategic marketing will be to determine which customers to approach and how best to approach them in order to execute a distinctive value proposition and build a long term customer relationship, in other words, to insure "micro-market management".

This assessment needs to be based on a detailed customer database which will be used across all the different channels, such as the direct sales force, call center, or website, and which allows each customer or potential customer to be managed individually across all the contact points he may choose. This very detailed information can then be used to support the sales force much more intensively than before. As a consequence, the sales force's approach and its success (or lack of it) will become much more transparent. Moreover, the function will also become much stronger, since it will have a much deeper understanding of its customers and will be able to offer those customers not just a product, but a convincing solution.

Setting the marketing and sales organization on a growth path in this way makes it critically important to insure that there is enough capacity along the whole order generation process to support that growth. This requires better understanding and better management of any resources that might prove to be bottlenecks. In businesses driven by customized solutions, these bottlenecks often lie in the technical service laboratories that are responsible for generating customized solutions (by suitably adapting basic formulae or by troubleshooting). Here, it is critically important to understand the capacity that is needed to support a growth based on some intelligent assumptions about the "hit" rate of sales pitches. It is also crucial to steer critical knowledge resources in a direction determined by the value of the work to be carried out for the customer and to remove bottlenecks by developing cross-functional capabilities (Fig. 13.4).

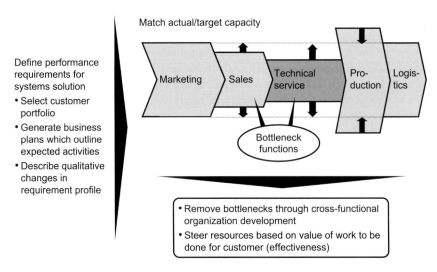

Fig. 13.4 Managing bottlenecks in the order generating process
Source: McKinsey

13.4
Securing Implementation by Superior Micro-Market Management

The change to customer-centric marketing has to be tightly organized to insure that the crucial knowledge needed to carry out the strategy successfully really is available and that the company has a very clear picture of how it will be applied. The transformation should be carried out in four steps, accompanied by the generation of the comprehensive customer database and an appropriate "e-collaboration" tool mentioned above, which will provide the basis for the fourth and final step.

First, a fresh analysis of market potential should be made to determine where the company is delivering well at present and to identify gaps which it ought to fill. In this step, targets will be set, ultimately based on the shareholder value which the marketing and sales functions expect to add to their company but which at the sales and marketing level translates into the growth and profitability of the customer portfolio. In the second step, the key dimensions that influence customers' purchasing decisions – for example, response time, delivery accuracy or the results of first tests – and the differentiating performance levels have to be understood and the market has to be segmented accordingly. This is the stage at which the target customer portfolio described in Section 13.1 should be defined.

In the third step, the organization needs to develop the customized offerings discussed above and also needs to make sure that all the conditions for their delivery are in place – for example, that all bottlenecks have been removed as described above in Section 13.3. In the fourth (and final) step, a broad process is in-

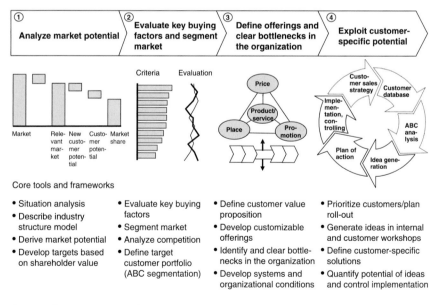

Fig. 13.5 Superior micro-market management process
Source: McKinsey

itiated to maximize each customer's specific potential, based on ideas generated in workshops both with the company's own staff and with the clients, and using the new customer database as its foundation (Fig. 13.5).

In conclusion, it seems that the marketing and sales functions need not fade away in the chemical industry as a result of commoditization and increased customer sophistication. If they move towards a more customer-centric strategy, they can in fact play a more important role than ever in supporting their companies' value proposition. Furthermore, the power of the organic growth these functions can generate should not be underestimated: five years of 15 percent annual organic growth, for example, would double the size of a company, rivaling a successfully implemented "mega-merger of equals".

On the basis of the three key changes we propose above, we have prepared a short checklist so that companies can diagnose their position in these areas (Fig. 13.6). For those who fall short, survival will depend on their ability to change.

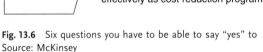

Customer strategy
- Are you growing at the same rate as your best customers?
- Do you have a list of customers you do not want to serve?

Management of product/ service mix
- Do you know what prevents your customers from switching?
- Can you serve all potential customers with customized solutions?

Sales management approach
- Is your growth free of resource constraints?
- Can you control sales growth as effectively as cost reduction programs?

Fig. 13.6 Six questions you have to be able to say "yes" to
Source: McKinsey

14

The Role of Mergers and Acquisitions

Markus Aschauer, Christophe de Mahieu, Philip Eykerman, Gary A. Farha,
Michael Graham, and Thomas Röthel

In a globalizing market with increasing competitive pressure, consolidation through mergers, acquisitions, and alliances can, at times, offer substantial benefits. Experience indicates, however, that integration at the business unit level can have advantages over full-scale mergers. Using the example of the European petrochemical industry, where there is still more scope for consolidation than in the USA, this chapter discusses the drivers for finding suitable partners, various types of deals, and their potential benefits.

14.1
The Case in Favor of Consolidation

An increase in the number of worldwide players in many chemical segments, new low cost entrants, and increasing opportunities for global price arbitrage have increased the pressure on chemical companies, particularly those with a heavy exposure to the commodity chemicals such as olefines, plastics, and fertilizers, which represent approximately one third of all chemical sales.

The pressure is particularly high in Europe and Asia, where most chemical subsectors are fragmented, leading to recurrent overcapacity and price wars. Europe, for example, has 15 or more companies competing in such markets as polystyrene and polyethylene. Structural weaknesses, such as subscale, non-integrated assets and the high costs of naphtha-based feedstock, make matters worse. Sixty percent of Europe's high-density polyethylene plants are small by world standards. In addition, many of them are not integrated with a competitive source of feedstock and are therefore at a cost disadvantage. As a result, their production costs may be up to 50 percent higher than those of nearby Middle Eastern producers, who generally have world-scale, fully integrated sites and can use cheaper ethane feedstock.

Consolidation through mergers, acquisitions, and alliances can be an effective way to deal with these forces. Its benefits can be twofold: cost reduction through synergies and economies of scale, and improvements in the structure and performance of an entire industry segment.

Our experience indicates that the savings realized by mergers range from 4 to 9 percent of the acquired company's sales in commodity segments and from 8 to 17 percent in specialty segments when the post-merger phase is managed properly. This represents an improvement in operating margins of 30 to 70 percent in commodity segments, and 30 to 60 percent in specialties.

The benefits of consolidation can include superior capacity utilization, the sharing of best practice process improvements, the elimination of overlaps in the distribution system, and the use of geographically closer plants for sourcing. Research and development and sales and marketing expenses can also be cut back by eliminating duplications in operations. Some non-traditional players (e.g., financial investors) have been able to slash their sales, general and administration costs by as much as 40 percent on the back of consolidation moves.

Undoubtedly, some of the savings could have been achieved by other means, but the act of combination can often serve as a catalyst in realizing value creation opportunities that might otherwise have remained latent.

Consolidation can also improve the structure and performance of an entire industry segment. In an analysis of a large set of commodity and specialty products, we assessed for each segment the likelihood that capacity rationalization benefits would be realized, the value they represented, and the feasibility of consolidation (Fig. 14.1). Improvements to industry structure appeared to be feasible and also to offer a source of considerable potential value for large product segments such as polystyrene, ABS, PVC, and polyethylene (Fig. 14.2).

In commodity segments, consolidation can lead to better decisions by individual firms about capacity additions (and hence to lower overcapacity), and can thereby

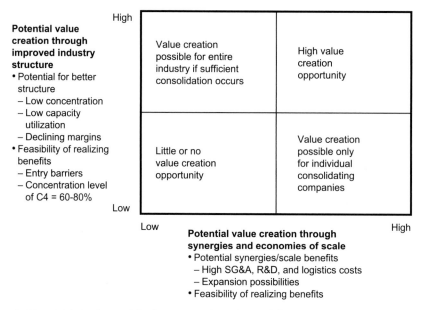

Fig. 14.1 Analysis of potential value creation through consolidation
Source: McKinsey

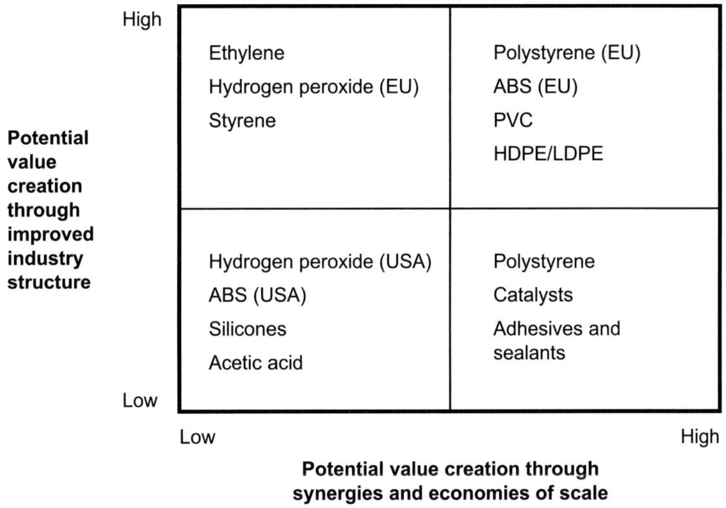

High

Ethylene Hydrogen peroxide (EU) Styrene	Polystyrene (EU) ABS (EU) PVC HDPE/LDPE
Hydrogen peroxide (USA) ABS (USA) Silicones Acetic acid	Polystyrene Catalysts Adhesives and sealants

Potential value creation through improved industry structure

Low

Low   High

**Potential value creation through
synergies and economies of scale**

Fig. 14.2 Value creation potential of selected chemical products
Source: McKinsey

reduce cyclicality. In specialty segments, consolidation typically extends the life of product differentiation advantages and value-based rather than commodity pricing.

These structural benefits can be difficult to achieve and particularly difficult to sustain, however. The conditions must be right: the overall industry must not behave destructively, and there must be no external shocks or discontinuities in the form of new entrants or new technologies. Companies seeking to assess the feasibility of creating and capturing such benefits must therefore also consider how the dynamics of the marketplace may change over time, and how long an improved industry structure can be sustained.

We believe that consolidation holds out greater promise in more fragmented markets, such as Western Europe, than in the United States (Fig. 14.3). It may be harder to unlock the value in Europe, but the rewards for creative firms that find a way will be all the greater. Several mergers in focused product segments have already taken place, including DyStar and Syngenta in specialty chemicals, and EVC and Basell in commodity chemicals. Depending on the segment, as many as 20 additional deals among medium-sized players are still possible in the European petrochemical sector before serious antitrust concerns are likely to be raised (Fig. 14.4).

The benefits of an improved industry structure can be compelling in a broad range of commodity and specialty segments. We estimate that reductions in overcapacity alone could allow a representative chemical company to create additional shareholder value equivalent to at least 5 percent of its current market value or about 10 percent of its invested capital base. When these figures are combined with the value that might be generated through synergies, the total value of consolidation for a typical chemical company comes to 10 percent of its market value or 20 percent of its invested capital base.

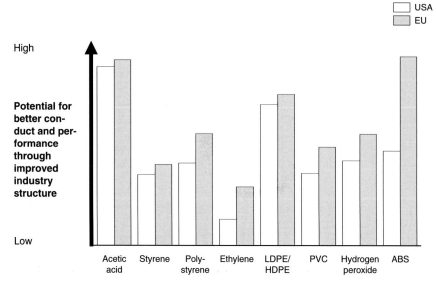

Fig. 14.3 Comparison of industry structure – degree of concentration
Source: TECNON, McKinsey analysis

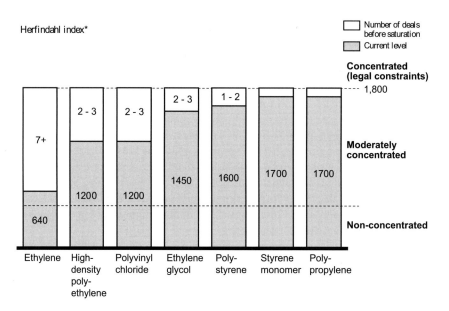

Fig. 14.4 Petrochemical sector concentration, Europe 2000
* EU Council's official criterion for defining market concentration (Herfindahl index: sum of squares of market shares of all players). Maximum market share of any single player has to be less than 30%; Source: TECNON, McKinsey analysis

Companies often fear the problems of consolidation, however: high acquisition premiums, the difficulty of finding willing buyers or sellers, and increased exposure to segment downturns. Such concerns can be difficult to overcome, and certainly the value derived from a transaction does not always cover the premium paid. All the same, companies that succeed at consolidation are often able to create substantial value over and above the merger or acquisition premium. Moreover, some particularly skilled players have used judicious timing or creative structures – joint ventures, asset swaps, or counter-cyclical acquisitions – to reduce or eliminate these premiums. Chemical companies should also consider timing their transactions without regard to their own situation, acquiring an asset when it is cheap on the market rather than at the precise moment when they need the capacity to grow.

14.2
Finding a Partner

If mergers, acquisitions, and alliances are the means to renewed profitability in the European petrochemical industry, how should companies go about choosing partners? A clear understanding of the value-creating potential of alternative deal options should underlie any decision – and narrow the strategic options. We estimate that, in many European chemical subsectors, no more than 20 percent of all the remaining possible deals are truly attractive.

Petrochemical companies with a number of businesses should first evaluate deal options for each of them. Most costs are incurred at the business unit level, so that is where to look for value-creating synergies. Big mergers can be found at the corporate level but are less likely to create value. In one recent petrochemical industry merger, for example, synergies were found on the oil side but virtually none on the chemical side, because the partners' businesses did not overlap.

Deals at the business unit level have another advantage, too: they reduce the potential for conflict over corporate control, a point especially relevant for small or medium-sized companies that fear losing decision-making power by merging with larger conglomerates. A deal with a rival closer to their own size may reduce this concern.

Companies can assess the potential value of any deal by considering how much scope they have to capitalize on five important value drivers, which are described below.

A rationalized asset base. By joining forces, some players may succeed in closing high-cost production lines and capturing the associated savings without sacrificing market share. The larger size of the resulting entity may give the partners enough idle capacity to transfer production from their least efficient lines to more efficient ones, which would then run at a higher capacity utilization. One or more of the remaining lines may have to be "de bottlenecked", or expanded at low cost, to accommodate the production volume that has been transferred from closed lines.

In other cases, joining forces may permit the partners to replace a number of sub-scale units with a single new world-scale one running at much lower cost.

Such units should be located within large chemical complexes (ensuring lower feedstock and operating costs) and be close to a port.

Potential partners should evaluate opportunities to rationalize their asset bases by identifying the improvements each of them could carry out by itself, and then identifying the additional improvements that could be achieved only by joining forces. The latter represent the true synergies accessible through a transaction.

In looking for production lines to close, companies should consider several constraints. It may, for instance, be hard to close lines that are physically integrated with feedstock, so optimum candidates for closure either will not be integrated, or will be linked to feedstock pipelines that would permit alternative uses of the feedstock should they be closed.

It may also be impossible to transfer production between lines using different technologies, because the products of these lines may be highly specific: emulsion polyvinyl chloride, for example, which is used for applications such as carpet backing, cannot be made on more common suspension polyvinyl chloride lines.

Finally, it may not be wise to close some higher-cost lines. They may well be devoted to high value-added products and thus operate at above average margins.

An improved cost position. Technological innovation and operational improvements mean that gross margins for petrochemical products will probably deteriorate over the long term, leading to a relentless price-cost squeeze. It is therefore very important for companies to improve their cost positions. This is particularly true for European players, which face increasing volumes of imports from the Middle East – imports that could threaten the viability of one-third of the European asset base. The history of most petrochemical sub-sectors shows that only players with first- or second-quartile cost positions have covered their cost of capital across the business cycle. This situation will not improve in the future.

Operational synergies. Mergers, acquisitions, and alliances – in petrochemicals as in other industries – will generate opportunities to cut operating expenses in sales and marketing, distribution, purchasing, administration, and research and development. Synergies in the first four of these categories (which account for the bulk of potential cost savings that are not related to fixed assets) will probably be highest for companies operating in the same geographic areas.

Such companies may share a number of customers, and this could create opportunities to cut sales, marketing and distribution costs, even for players in different sub-sectors, since many plastics have partially overlapping applications and can be sold to the same customers.

Companies in the same areas usually also have better opportunities to reduce their purchasing costs: because of high transport expenses, a number of important raw materials, such as olefines, are typically purchased locally, and their producers tend to be geographically fragmented.

While it is difficult for the outsider to estimate the full extent of these synergies, potential partners should at least be able to assess their likely order of magnitude. If more than half of the capacity of the partners is located in the same country, they can probably achieve substantial operating synergies. If more than half of it is located in adjacent countries, potential synergies are likely to be smaller, and they will be smaller still if most of it is located in countries that are not adjacent. Of course, potential partners may succeed in reducing the cost of maintaining their corporate headquarters and research and development programs (from 1 to 3 percent of return on sales) regardless of their geographic locations.

Transfer of technology and skills. Petrochemical players can make themselves more cost competitive and increase their market penetration by focusing on technology and operating skills. A leading-edge process technology (e.g., Union Carbide's Unipol polyethylene process) can have a cost advantage of more than 10 percent over traditional process technologies. Such leading-edge products as metallocene polyethylene and polypropylene – chemicals that make it possible to manufacture better quality end products at a lower cost – can revolutionize markets and give their owners a strong competitive advantage.

Deals can create value on the technology side if they unite players with different (and, ideally, complementary) skills and strengths. That has been the rationale behind a number of recent technology-focused alliances, such as the one between Union Carbide and Exxon in polyolefines: the former excels in process technology, the latter in metallocene technology. Clear synergies exist, since a producer can tailor a process to the specific needs and opportunities of new metallocene products.

Sharing manufacturing best practices and internal benchmarks can be a major source of value, especially when a deal unites players with widely different levels of skill. A typical deal of this sort would be a merger between a large global player and a smaller one, perhaps controlled, until recently, by the state.

Vertical fit between partners. Olefines, such as ethylene and propylene, are the key raw materials for most petrochemical products. No fully efficient and liquid merchant markets for olefines exist yet in Europe, because of the traditional integration of petrochemical players and also because of transportation difficulties, which are worsened by the absence in Europe of an extensive pipeline network like that on the Gulf Coast of the United States. Such markets probably will not be developed in the short to medium term, especially for ethylene which is relatively hard to transport.

To reduce their sourcing risks, European petrochemical companies may thus benefit from some degree of backward integration into olefines. Players with limited olefine production capacity are probably in the best possible position: they can secure supplies while avoiding the temptation to undercut prices of downstream products during downturns. Players with an excess of olefines may be exposed to this temptation in order to maximize the capacity utilization of their olefine production, where fixed costs tend to be high.

However, increased integration into downstream businesses like plastics processing is likely to be less valuable, for downstream markets tend to be very competitive and require an understanding of consumer markets. Also, taking a downstream position that involves expansion into a customer's business puts companies in a potentially uncomfortable competitive position.

14.3
Is there an Alternative?

Sometimes the right partner for a merger, acquisition, or alliance just does not exist, or it exists but its size is intimidating. As mentioned before, some medium-sized and small players are reluctant to enter into potentially attractive deals with their larger competitors for fear of losing control over their affairs. In the absence of any truly attractive alternative, the temptation for them is to pursue leadership in their regions or customer groups of choice.

While it is outside the scope of this chapter to discuss the virtues and problems of such strategies, they could obscure substantial, now-or-never value-creating opportunities. If most of the industry consolidates, capturing the available synergies, the competitive position of the remainder may be substantially weakened.

To avoid this situation while preserving corporate control, small or medium-sized players could consider smaller deals with partners of a similar size, followed by a larger deal with a leading player. This is not a simple solution, and some time may be necessary between the two deals because of the difficulties of post-merger management (see Chapter 15).

Another strategy is to negotiate alliances or deals below the business unit level – arrangements that concentrate on a single asset or a single stage of the business system. An alliance might, for example, involve an agreement to share logistics with another player on a single site. A deal could involve the takeover of one plant from another company.

There are many possibilities, though their value-creating potential is much smaller than for deals at the business unit level because there are fewer synergies: in polyolefines, for example, traditional agreements and alliances below the business unit level can achieve no more than 30 percent of the full potential for improvement. Moreover, organizing a set of deals with a number of heterogeneous partners may prove extraordinarily complex.

Nonetheless, and subject to anti-trust considerations, some types of deal below the level of business units could be attractive, particularly to petrochemical players. Such alliances or deals may also appeal to potential partners who would like to get to know each other better before they enter into a full merger, or to players who lack exciting merger, acquisition, or alliance options at the business unit level. Three examples are given below:

1. Improving plant specialization and swapping products. An alliance for joint production planning can improve the specialization of the partners' plants and therefore promote longer production runs and increased output for the parties involved. Such an alliance would certainly have to include an agreement to swap products as needed. For high-density polyethylene players, this approach could lead to improvements in the order of a 1 percent return on sales.

An alliance of this type poses certain dangers, however. For one thing, its scope may be limited by the desire of one or both partners to protect proprietary know-how. In the second place, arranging product swaps can be complex and may require repeating a lengthy testing process with some customers a number of times.

2. Orchestrating freight and packaging. Players who share a single site may get together to outsource their packaging, storage, and distribution activities to third party service providers. This opportunity is also available to non-contiguous plants. Through standardized design and engineering, as well as superior execution, specialized service providers may have an advantage over petrochemical firms in developing a logistics platform. This advantage can be amplified when a number of customers share storage facilities – an arrangement that makes it possible to capture substantial economies of scale. The efficiency of shared storage facilities is so great that the service provider may be able to pass on to the petrochemical companies savings as high as a 1 percent return on sales.

3. Participating in the development of a "condo cracker". Building a world-scale cracker can cost about USD 1 billion. No more than two new world-scale crackers will probably be needed to satisfy the increase in European demand for raw materials over the next eight to ten years. European petrochemical players could therefore join forces to build "condo" crackers owned by a number of them, each with the right to use or sell a fixed amount of a product at a predetermined cash cost.

Such condo crackers would allow the industry to maintain an even balance of supply and demand in Europe, while permitting the partners to secure access to cheap feedstocks. Compared with the alternative – the proliferation of small crackers, leading to higher production costs and overcapacity – the benefit would be substantial for the whole sector.

This is a difficult time for Europe's petrochemical industry, but there is hope. Mergers, acquisitions, and alliances can give some players a unique opportunity to improve their fortunes. The number of attractive deals, however, is limited, and in many sub-sectors no more than two or three large deals could be consumated before antitrust limits are reached. Players that cannot or will not contract deals at the business unit level can find alternatives that will let them retain corporate control while extracting some, though not all, of the economic benefits of more sweeping deals.

15

The Delicate Game of Post-merger Management

Tomas Koch, André M. Schmidt, and Boris Gorella

The high value creation potential of mergers and acquisitions often proves hard to realize in practice, particularly in the case of very large transactions (Chapter 14). According to most estimates, fewer than 50 percent of mergers and acquisitions ultimately add value, and, as often as not, the strategic liabilities arising from the deal turn out to be greater than the synergies.

The chances of success can be increased, however, by effective post-merger integration. In essence, this enables the new entity to realize all the expected synergies over the longer term, and also to identify new potential.

A merger is a new beginning, giving the resulting company what seems to be the ideal opportunity to rethink its entire approach to its business. While future strategy and the organizational structure within which it will be implemented are extremely important, arguably at least equally crucial to post-merger success is the establishment of a joint corporate culture – an essential prerequisite for getting all employees to "paddle in the same direction", which often proves harder to achieve than expected.

In this chapter we discuss what happens in the case of a full merger; several of the elements considered will, however, also apply to lesser levels of integration.

15.1
The Obligation to Create Value

The ultimate measure of success is when the merged company creates more value than the separate entities from which it was created. Unfortunately, these expected synergies do not all accrue to the new entity: the original owner of the target company also has to have its share. Worse still, that share is usually the largest part of the expected synergies, transferred directly to the seller in the form of a premium paid by the acquirer. It is the acquirer's obligation not only to realize the premium paid, but also to add further value in the form of synergy gains.

Value creation can be estimated by the development of the total return to shareholders, in terms of both the immediate reaction of shareholders in the first 5 days after the announcement and the long term development of a stock 2 to 3 years after integration. The first of these mirrors the financial markets' expectations of the new company, the latter their realization. Interestingly, as we saw in Chapter 2, financial markets' expectations have proved very reliable; if markets err, they err on the optimistic side. Fig. 15.1 shows the percentage change in total return to shareholders of the acquirer relative to its index two to three years before and after announcement of the merger (value realization) contrasted with the percentage change in total return to shareholders five days before and after the merger announcement for the same stock (expected value creation).

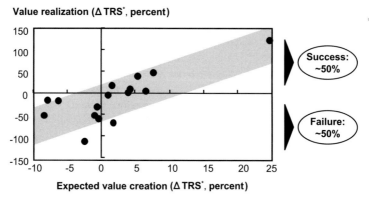

Value realization (Δ TRS*, percent)

Expected value creation (Δ TRS*, percent)

Fig. 15.1 Analysis of long term value realization versus value creation expected by the financial markets
* Change in total return to shareholders relative to index two to three years before and after announcement of the merger; ** change in TRS performance five days before and after announcement of the merger; Source: McKinsey

USD millions

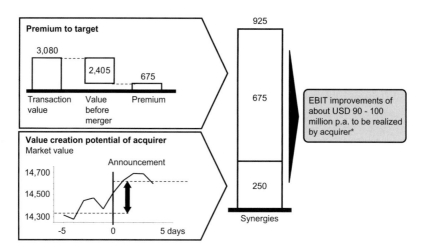

Fig. 15.2 Expectations of the financial markets on total synergies
* Assumption: WACC of 10%, no growth; Source: McKinsey

The premium that needs to be covered can be pretty hefty. The company in Fig. 15.2, for example, obliges the acquirer to realize total synergies of at least USD 925 million to sustain its market capitalization in the long run, or annual synergies of USD 90–100 million.

The top management of a merged entity is therefore faced with major value generation challenges in the long term. It should, however, set itself even higher targets in order to surprise the financial markets and create even more value. Practically speaking, this means that the top management must foresee the market reaction by assessing the obvious synergies subtracted from the known premium and also have further ideas for value creation on hand.

15.2
The Keys to Success

To address the problem of successful value creation, we performed a number of in-depth analyses of mergers and acquisitions in the chemical industry. We analyzed the top 25 mergers and acquisitions based on transaction value in the last ten years outside-in, and then interviewed most of the CEOs or integration leaders of these projects to find out how they had led and structured the process. The results show that success essentially depends on three factors (Fig. 15.3).

Decisive leadership. This means a management team that is ready at all times to make clear and rapid decisions on fundamental issues such as organizational structure, the definition of areas of responsibility, or the selection of personnel.

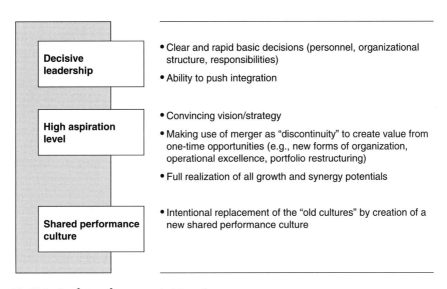

Fig. 15.3 Key factors for success in integration
Source: McKinsey

Corporate managers have to systematically promote integration and create and preserve the conditions required for a successful merger.

A high level of aspiration. Since successful integration is measured in terms of financial performance, it is crucial for the new company to set ambitious financial goals for itself from the very beginning, to force it to take full advantage of all the available synergies. The key gains in this situation will come from cost reductions along the entire value chain on the one hand, and from increased revenues due to the expanded customer base, knowledge transfer, and development of system solutions on the other.

A shared performance culture. Merging companies should also look upon the integration process as a unique opportunity for reorientation, for example, completely reorganizing their areas of business or introducing new incentive systems. They should also look to develop a new and inspiring vision and a strategy for growth.

For example, when German specialty chemical companies Degussa-Hüls and SKW Trostberg merge they propose to form the world's largest specialty chemical company, with an attractive portfolio. In addition to organic growth, divestments of non-core businesses in the range of EUR 5 billion and acquisitions of EUR 10–12 billion are planned. With about EUR 18–20 billion in sales, the new company (Degussa) will be the global leader in specialties.

The merging of corporate cultures is one of the most difficult and neglected areas of integration. Most companies still concentrate on systems and structures at the expense of the human elements. In order to foster the full cooperation of all the personnel in a merged entity, a new company-wide performance culture must be agreed upon. For example, in one of the biggest mergers in polyolefines one company was used to giving itself targets from a stringent top-down perspective, whereas the other tended to work very much in a bottom-up style.

Matters inevitably become even more difficult when fundamental cultural differences are compounded by linguistic ones. This was a particular challenge, for example, in the formation of Aventis from the merger of Hoechst and Rhône-Poulenc's pharmaceuticals and agrochemicals activities.

However, even without language differences, distinct management cultures can create difficulties. Dow, for example, with its strongly market-oriented perspective, had to place great emphasis on creating a new culture when it made plans to integrate with Union Carbide, another American company which was mainly science- and technology-oriented.

Thus, for companies with strongly contrasting corporate cultures, a new and shared performance culture has to be actively created (Fig. 15.4). Performance cultures can be measured in two dimensions. The first of these assesses the "quality of direction". If it is high, there is considerable consistency in the objectives of the group's individuals, and the objectives are challenging. Indicators are usually a common vision, a well-defined strategy, and aspirational targets.

The second dimension assesses the "quality of interaction". A high rating here means that the individuals in a group are very supportive in their interactions, with a high degree of mutual understanding. The interaction processes contain few redundancies and are highly focused. People in groups like this are usually highly motivated.

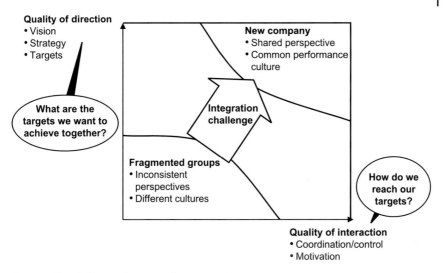

Fig. 15.4 The challenge of a new performance culture
Source: McKinsey

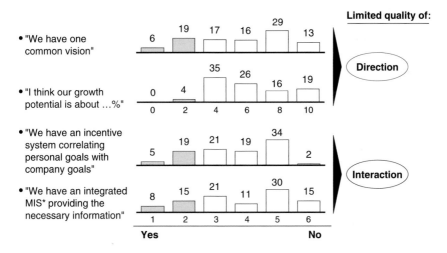

Fig. 15.5 Results of a management survey
* Management information system; Source: McKinsey

Integration strategies have traditionally amounted to little more than the definition of new corporate structures, rules, and processes, formulated in the hope that the two corporate entities would simply merge by themselves in the fullness of time. These strategies, however, failed to consider that each of the individuals involved has his or her own personal goals, experience, values, and business interests. It takes more than putting words down on paper to make the management team and other employees believe in the new goals, far less personally pursue and support them.

At one merged company, the management group of a newly formed business division was surveyed about their objectives and way of interacting. The level of agreement on most questions was below 25 percent (compared to best practice companies with typically more than 80 percent consensus), and the generally wide distribution of the different answers clearly showed that the group lacked consensus in several areas, including the crucial question of a common vision (Fig. 15.5).

Firms have to take into account the sharply contrasting mindsets of different groups of employees. Managers from the world of e-commerce, for example, differ greatly from those formerly employed by state-owned organizations. Yet such people increasingly find themselves having to work together in teams, often as the consequence of a merger. It is important to try to unify their diverse points of view and, in doing so, to develop a joint, shared performance culture. This is particularly vital for the top management team, which not only takes global strategic decisions, but also sets a highly visible example for the rest of the organization.

15.3
The Road to Successful Integration

The integration process itself can be long and involved, and can easily take up to two years, depending on the scope and complexity of the merger. It is often broken up into a large number of small individual projects, and to make it all work together it has to be backed by a well-structured project management plan. Such plans can be effectively divided into two phases (Fig. 15.6):

Phase one: integration design. In this phase, all the goals to be achieved by the integration (and the means of achieving them) have to be clearly defined. Some background conditions for success are also laid down at this stage, such as scheduling. Despite its relatively brief duration of 2 to 3 months, this phase is disproportionately vital to the success of the process, because all the essential decisions defining the new company are made as it progresses. Experience shows that problems arising in the second phase usually stem from an incomplete design phase. The integration design is usually prepared by the top management.

Phase two: actual integration. Over a period of 1 to 2 years, a large number of project teams work on various topics, coordinated by an integration office. We will see later how to organize this effectively. Properly handled, the integration phase should mobilize the entire company.

SIMPLIFIED

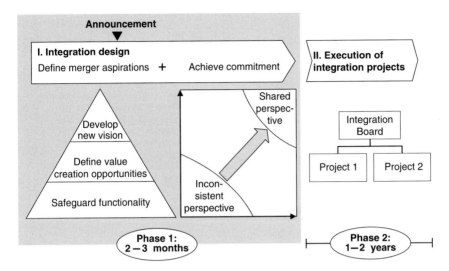

Fig. 15.6 Integration process
Source: McKinsey

15.3.1
Phase One: Integration Design

The main task in the first phase is to set goals at a suitably challenging level which has the backing of the entire new management team – a team that must be reasonably homogeneous and totally dedicated to the new company. This mindset must be embryonically present from the start, and will become increasingly established as the work goes on.

Four aspects of the goals are particularly important, and should be taken into consideration as early as possible in the planning stage: developing the vision, identifying possible ways to add value, insuring the efficient functioning of the new organization, and creating a shared performance culture. The second last of these is particularly crucial if delays in implementation are to be avoided and the negative effects on the company's current business activities kept to a minimum.

The measures to be taken along these three dimensions can be divided into general goals and those for which industry knowledge is required (which we have called "industry-specific"). We used this split in order to show which drivers are universally valid in a form that can be applied in any industry and which drivers have to be adapted specifically to a given industry (Fig. 15.7).

	General actions	Chemical industry–specific actions
Vision		• Refine core competence based vision • Develop motivating growth strategy based on vision
Value creation	• Introduce value based management and KPI systems • Introduce incentive systems	• Design infrastructure company to enable portfolio management and bundle technical competence • Develop long term site strategy • Introduce "best-of-best" benchmarking for sites/plants • Change market approach • Capture direct merger synergies
Functionality	• Stabilize current business • Design corporate structure • Select management • Develop joint HR strategy • Align IT systems • Develop communication strategy	

Fig. 15.7 General and industry-specific integration measures
Source: McKinsey

15.3.1.1 Developing a Shared Vision

The new company's future platform has to be announced as soon as possible for two reasons: first, a clear vision prevents uncertainty in the organization and secures the motivation of the employees. This is especially important in today's War for Talent (see Chapter 10). Second, it serves as a proactive demonstration to the financial markets that the management has the firm intention of reorienting the new company, so that shareholders respond to the company's targets rather than setting their own goals.

Merging companies have to make sharp decisions on what they want to regard as their core competences, as a basis for the new growth platform. Non-core business areas need to be identified early on, since it will be absolutely essential for the new company to clear them out of the way of the new development. Degussa-Hüls and SKW Trostberg once more serve as a good example: formed into the world's largest specialty chemical company by owners which are not mainly chemical companies, the new entity will have to divest several divisions originating from the former Hüls and Degussa.

15.3.1.2 Identifying the Potential for Adding Value

Since the success of a merger is ultimately measured in terms of the value it generates, synergies that were estimated in broad outlines when the merger was planned have to be specified in detail when both parties suddenly start playing with totally open cards. Two questions in particular must be addressed when formulating overall goals:

1. Will the synergies that emerge be sufficient to meet the financial markets' expectations?
2. Are all the possibilities for adding value being taken into account?

The overall goal can be derived in one of two ways: it can either be based on the estimates from the due diligence – a "top-down" goal set by the companies' management – or it can be built from the "bottom up" by the individual project teams. A pure example of either case is very rare in practice; usually, a weak form of the top-down variety prevails, in which the company's management formulates objectives in cooperation with the relevant project staff.

The potential for adding value can be sought in two main areas: the business itself, and the new leadership and management systems. The directly business-related potential is likely to come from the following actions:

- Reinforcing purchasing strengths by selecting the most cost efficient suppliers. This can be further enhanced by realizing the quantity discounts that will become available because of the company's increased volumes and its redesigned purchasing processes. This will apply to almost all mergers and acquisitions in which product overlaps exist.
- Improving production and technology by developing a long term site strategy, and improving the production facilities by the transfer of expertise and the introduction of fresh management thinking. One plastics producer, for example, started a benchmarking competition among all its sites and "built" a theoretical best-of-best plant on the basis of the information obtained. It was thus able to improve productivity by giving all plants specific goals and creating in-house competition.
- Reinforcing marketing and sales by tapping into cross-selling opportunities, a newly developed systems business, knowledge exchange in the technical service function, and the consolidation of sales organizations. Bayer completed its polyurethane business by the acquisition of Lyondell's polyol. Through the combination of isocyanates with polyols, Bayer is now well positioned to provide full system solutions.
- Introducing new service concepts, giving the business units the opportunity to focus on their core competences, and additionally introducing a stronger meritocracy into the newly formed service centers. When Degussa and Hüls merged, they strengthened the position of Hüls' young service company Infracor to focus further on their core competences.
- Trimming down the management team by precisely defining the duties and competences of the organization's central functions. At the plastics producer mentioned above, the management team was reduced to its main functions by means of maximum decentralization and project-based organization. In addition, the management moved to a new location which had to be built up from scratch, in order to create an entrepreneurial environment.

15.3.1.3 **Securing Current and Future Business**

The most important task here is to insure that the new company becomes functional as quickly and as smoothly as possible. It is, therefore, essential to decide as soon as possible what has to be done, how it has to be done, and who is responsible. Thus, the new organizational form must be selected at a very early stage, management personnel has to be selected fairly and quickly, and responsibilities must be allocated to the relevant managers selected.

Choosing a new organizational form is not a problem when the companies involved already have a similar structure. But where they differ, a single structure has to be chosen for the new business, and that is not always easy.

When Hüls merged with Degussa, Hüls was being run as a slim chemical holding group with 12 independent companies, whereas Degussa was running its three business segments under a single parent company supported by very strong corporate center divisions. During the merger, both structures were evaluated and it was decided to operate the new company (Degussa-Hüls) as a decentralized parent company with four business segments and one service unit.

When it comes to staffing, the entire executive/management board and the CEO – in other words, the top management level – are almost always named at the very beginning of the merger. The second and subsequent levels, however, are often more difficult to fill because the members of the executive/management board are usually only familiar with the employees from their own organizations.

Personnel selection can take place in one of two ways, depending on the management culture. One way is for the executive/management board to make top-down decisions. However, a structured selection process can also be used in which each company nominates two or three candidates for each position, and these candidates are then interviewed by their respective supervisors and by the human resources director. Very often, some human resource consultants are involved in the process as well. The evaluation is ideally performed along defined criteria and should be as objective as possible.

There is no one right answer either to the organizational structure or to the executive selection process of management. These elements depend very much on the cultures the two companies are accustomed to. Furthermore, the best organizational structure will vary depending on the future objectives of the business (see Chapter 10). Whatever decisions are taken, however, they have to be taken fast.

15.3.1.4 **Building the Shared Performance Culture**

The programs and concepts described above only relate to the formal and substantive aspects of the merger. They do not take into account the personal perspectives of the employees affected. Nevertheless, incorporating these widely contrasting individual perspectives and, at the same time, reaching a reasonably high level of consensus is one of the most important achievements of the design phase. During this process, the future style of interaction between all employees should also be defined.

Concepts, after all, are only as good as the people who execute them. Their success depends in essence on three factors: first, the individuals' own goals and

Percent

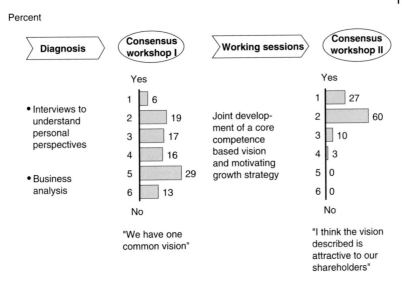

Fig. 15.8 Two-step process to develop a shared perspective; results of a management survey during two workshops
Source: McKinsey

their internal formulation of these goals; second, their interpretation of the existing opportunities and their freedom of action within the company; and third, their identification with the task in hand.

A two-step process has proved a valuable tool to address the individual perspective in building a shared performance culture. Fig. 15.8 shows an example of a division of a chemical company for which a common vision was to be developed.

In the first step, in-depth interviews are conducted to explore personal attitudes. In these, the independent interviewer should not focus on facts alone, but also on the very personal level of reasons and motivation. Typically, for example, the organizational structure of product areas is not defined by logic, but more often by a manager's personal desire to incorporate highly profitable areas into his own area of responsibility to mask weaknesses in his current portfolio.

By analyzing the interview results, an initial hypothesis can be formed about the strengths and weaknesses within the organization, and about those areas where the management team might have different attitudes to the goals and to cooperation. These points should then be explored further in workshops aimed at surfacing some initial ideas for solving the problems highlighted. The "performance ethic" framework discussed in Chapter 10 has proved very successful in surfacing issues.

The results of the first consensus workshop should be used to define projects which will then focus on finding firm solutions. In this second step, it is crucial to staff project teams with members from both of the pre-merger companies. After an intense working phase, each project team has to present its results to the whole management team in a final consensus workshop, giving the shared perspective of the new team.

Clearly, the process described is not sufficient to create a shared performance culture, a process which takes years of joint effort. However, it is the critical first step in defining shared directions and shared modes of interaction as the platform for future growth at the level of each individual employee.

15.3.2
Phase Two: Execution of Integration Projects

The projects needed to achieve the defined objectives start at the beginning of the second phase. In order to control the number and complexity of the individual projects, a separate organizational structure must be set up for the purpose, with clearly defined responsibilities and roles.

A rough outline of the organization for the integration project – for example, the establishment of an integration board and an integration office – should be drawn up as early as possible in the design phase. Refinement of the project's structure – that is, its separation into smaller projects and sub-projects – and the establishment of controls do not occur until the beginning of the second phase.

The design of the project portfolio (and also the allocation of resources) is based on the objectives that are determined by the management team. We have found it useful to rank each project – typical major ones include projects to develop the vision, to realize the full potential to add value, and to insure the effective operation of the new businesses, in line with the three key aspects described

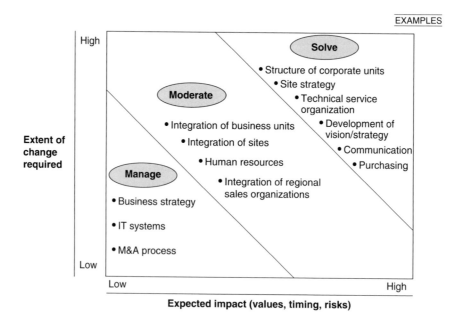

Fig. 15.9 Designing the project portfolio
Source: McKinsey

above – using a matrix defined by the axes "Extent of change required" and "Expected impact" (Fig. 15.9).

A three-layer project team structure has proven sufficient for smaller companies with a limited number of product groups and with revenues of around EUR 500 million. With the larger "mega-mergers", however, a four-layer structure works better. In the case of smaller companies, the organization consists of an integration board (the overall steering committee for the merger), an integration office (which encourages and supports the search for solutions to the various problems raised by the merger), and the project teams (which decide on the actions to be taken). With mega-mergers, the integration board should be further supported by dedicated steering committees which decide on the actions to be taken by the various sub-projects.

The integration board should consist of between two and four executive board members from the new company, the number depending on the scope of the integration. Its fundamental task is to control the general direction of the integration and to formulate guidelines for the individual project teams. It has to allocate the necessary human and financial resources, and to adopt measures that will bring about the declared objectives.

The steering committees (which are only required for mega-mergers) are specifically assigned to individual projects. Two executive/management board members are sufficient here, one of whom should be the person responsible for the new area in the future. In contrast with the weekly meetings of the integration board, the steering committee need meet only once every three to five weeks.

The integration office should consist of four to six staff members who have been freed up specifically to work on the project. One of them should be appointed as the director. If possible, the members should be managers with a good reputation in both organizations. The fundamental task of the integration office is to steer the whole process, moderate discussions in difficult situations, and set up sessions in which the project teams present their results.

The individual project teams are the "cells of the merger". They come up with ways of building effective operations that create value, as well as insuring that various areas such as research and development are aligned with the new organization's vision. Project teams should consist of two to six staff members, the exact number depending on the complexity and urgency of the project. Each project team leader should ideally be the future head of the division in question.

Controlling the integration project involves monitoring the progress that is being made towards the achievement of the overall goals. It also involves keeping an eye on the individual project teams as they move to fulfill the master plan for the merger. The intranets available today in most companies make it possible as well as desirable to broadcast the state of play on a regular basis to selected groups of people within the organization.

The integration office should actively support the teams in problem solving where projects are expected to have a high impact and imply a considerable degree of change. All the other projects can be supported as and when required. This may mean, for example, actively moderating the process of problem solving while taking care not to provide the solution. In most cases, however, the project teams can be left to work on their own. In these cases, the only duty of the integration office is to supervise the goal achievement process based on defined targets.

Merging two companies successfully and achieving a positive culture change is a major undertaking and a huge challenge for everyone involved, employees as well as top managers. In order to get to grips quickly with this challenge, the following ten rules should be meticulously obeyed:

1. Secure functionality and the ongoing business
2. Focus on value creation
3. Develop a new shared vision
4. Develop a new shared performance culture
5. Utilize the merger as a unique opportunity to revive the company
6. Exploit the momentum in the first 90 days after the announcement
7. Put speed before perfection
8. Do not make personnel concessions at the cost of content
9. Make top management drive the integration process
10. Communicate openly and quickly.

16
Cyclicality: Trying to Manage the Unmanageable
Paul Butler, Robert Berendes, and Brian Elliott

Producers of many commodity chemicals such as styrene, para-xylene and polyole-fines have long been plagued by huge swings in prices (Fig. 16.1) and operating margins (Fig. 16.2). The phenomenon is, of course, well known in the industry: during a cyclical fly-up, return on invested capital (ROIC) can exceed 60 percent. In the ensuing trough, ROIC may fall to below a company's cost of capital for prolonged periods – sometimes continuing for several years.

Such cyclicality causes great difficulties for managers making strategic and operational decisions, though interestingly enough the average total shareholder returns over a whole cycle are similar to those in non-cyclical industries. In a typical cyclical commodity, 70 to 75 percent of the operating margin is generated during the fly-ups, which last for only around 30 percent of the cycle, yet critical decisions about new investments have to be made during periods of poor, or even negative, returns. Ideally, business unit returns should be judged in the context of

USD/tonne

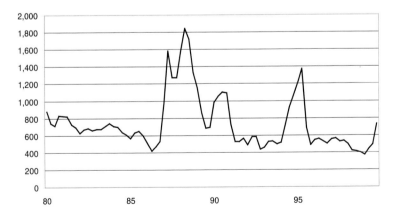

Fig. 16.1 Volatility of European styrene price since 1980
Source: TECNON, McKinsey analysis

USD/tonne

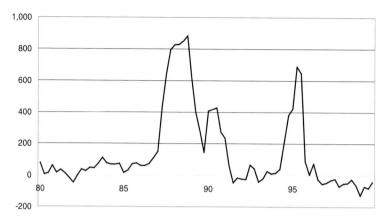

Fig. 16.2 Operating margin "fly-ups" in styrene since 1980
Source: TECNON, McKinsey analysis

a full cycle, lasting perhaps 8–10 years, but this is rarely if ever done, or is impractical for decision making purposes.

To make matters worse, it has proved to be difficult to predict cyclical peaks and troughs with any certainty. Some markets are so finely balanced that relatively small and unexpected events on either the supply side (such as plant breakdowns) or the demand side (such as higher- or lower-than-expected consumer spending) can send prices soaring or plunging with almost no warning. Even detailed microeconomic analysis aimed at projecting operating rates and margins cannot always pick up future price swings with certainty. Such analysis is also becoming more difficult as markets become more global and inter-regional product flows play a greater part in setting prices.

As a result of the above factors, finally, senior managers find it hard to explain the causes and consequences of cyclical prices to customers, to their corporate centers, to outside investors, to their employees – and even to themselves.

It is unrealistic to believe that any single company can control or eliminate cyclicality in any particular product sector. However, we have developed a model which makes it possible to understand and explain to some extent what drives and sustains cyclicality. Managers can then begin to understand the levers at their disposal to mitigate its worst effects, chiefly in the form of correct timing of capacity investments as well as endeavors to predict demand and also price in order to improve planning of investments and strategic presence in certain sectors. However, they should be careful to weigh up the alternatives – it is extremely difficult to use these levers to gain competitive advantage. Much greater and more certain benefits are to be gained from measures to improve operational performance.

16.1
The Drivers of Cyclicality

There are many theories about the causes of cyclicality. A view common both inside and outside the chemical industry is that imbalances in supply and demand occur because companies invest at the top of the cycle (when prices are high and funds are available), just when demand is about to tail off.

A second hypothesis is that firms which are thinking of adding new capacity are unable to forecast future demand with sufficient accuracy to ensure that prices are stable as their new supply comes on stream.

A third theory holds that price cyclicality occurs because the balance of supply and demand is upset when new capacity comes on stream in large lumps – as it often does, because of the need to exploit economies of scale – rather than gradually. In addition, a fourth view holds that companies mistime their investments because they are unsure of what the industry's capacity really is.

All four theories are plausible, but each has different implications for what companies might do to gain competitive advantage in a cyclical commodity industry. It would therefore clearly be helpful to know which (if any) of these effects has a dominant influence on the volatility of prices and margins. We built a business dynamics model to attempt to find out.

Many factors interact and ultimately determine the market prices of commodity chemicals – and hence returns to the various players. Some of these factors are reasonably "hard", for example, cost structures and historic demand. Others are quite "soft", for example, beliefs about future demand, beliefs about what and when new capacity may be added, and expectations about future price movements.

The interplay of this complex mix of hard and soft factors is best understood by applying a business dynamics approach. A business dynamics model can, for example, simulate the cost structure, economics and investment behavior of a group of players in a commodity industry. The output is the financial performance of the industry – and even of individual players – over time.

The model was built to make a detailed investigation of the size and structure of returns in one particular chemical sector (terephthalic acid) over a period of 60 years – from infancy to maturity – under a huge range of assumptions. The model was structured so that growth in demand requires the industry to make frequent decisions about the timing and size of new investments. Rules embedded in the model – such as the need to return cost of capital over the long term, to maintain an average spare capacity margin of 10 percent, and to limit the amount of debottlenecking that may take place – provide the background against which investment decisions are made. There are also rules governing how and when companies can expand new plants as operating rates rise, or mothball old ones when operating rates fall.

At the core of the model is a dynamic relationship between product price and the industry's operating rate, that is, its balance of supply and demand. It is the fact that this empirically observed relationship is non-linear that leads to short-term price fly-ups separated by long troughs in which, typically, few players make money.

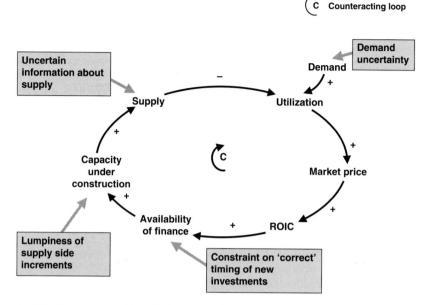

Fig. 16.3 Dynamic causal loop for investment decisions in the chemical industry
Source: McKinsey

Fig. 16.3 is a simplified dynamic loop showing the process by which investment decisions are made in commodity industries. It also indicates the points at which each hypothesis might have an effect on supply and/or demand, and hence on the operating rate.

By running the simulation model under various conditions it was found that three of the four hypotheses about the causes of cyclicality could be validated, with the first two being particularly powerful drivers. The fourth hypothesis – that poor information about other suppliers' capacity causes supply and demand to fall out of line – did not appear to be valid.

- *Investment at the top of the cycle.* We paid particular attention to being able to test this hypothesis. Modeling the relationship between the business unit and its corporate center is, we believe, especially important in understanding the psychological constraints on financing new investments when product prices are low. A typical situation in which a corporate center is unwilling to release investment funds during a price trough was simulated. The company's reluctance causes a delay in the building of new capacity, and contributes to a glut of new supply when it does finally come on stream. This "willingness" of a corporate center to finance continued investment in a cyclical business unit is expressed as a combination of recent returns on invested capital and the industry's expected future operating rate.

When investment was constrained during a price trough, price cyclicality in the resulting simulation was similar to that observed in actual chemical industries

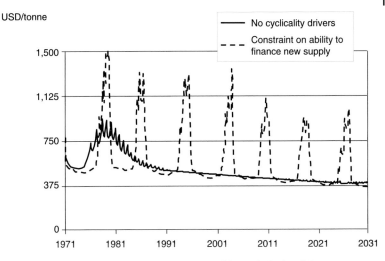

Fig. 16.4 Cyclicality generated by a business dynamics model in which the ability to invest at the "right" time is constrained by lack of finance
Source: McKinsey analysis

(Fig. 16.4). When the financing constraint was removed, and investment allowed at any time in the cycle, cyclicality disappeared.

- *Inability to forecast future demand.* When projected demand growth in the model was given even small perturbations, in line with medium term GDP cycles, realistic patterns of product price cyclicality were generated. The industry, as simulated, was unable to make accurate demand forecasts. Hence it persistently miscalculated the future balance of supply and demand. In reality, companies might be expected to learn about GDP cycles and time their investments accordingly. However, GDP cycles are themselves difficult to predict, and demand for chemical products is often extremely volatile regardless of GDP cycles. This can be because of stocking/destocking effects, for example, or because of a change in customers' usage of the product. We have concluded, therefore, that misjudging future demand can be an important factor which perpetuates cyclicality.

- *New capacity coming on stream in large lumps.* When new capacity was added in very large increments in the model, it was possible to generate reasonably realistic price cyclicality, but only if these increments were large compared with demand growth. This commonly happens during the early stages of an industry's growth, or when a product is mature and growth has slowed down. Hence, this driver may be relevant only at certain times in the lifetime of a product.

That said, there are others factors apart from product maturity that determine whether capacity increments are large compared with demand growth. In most chemical product sectors, minimum economic plant size increases quite rapidly over time as a result of technological innovation, and can lead to very lumpy supply-side additions. On the other hand, when industries become extremely large, annual volume growth of no more than 3 or 4 percent may make world-scale supply-side increments look modest.

- *Companies' uncertainty about industry capacity.* We found that realistic price cyclicality could be generated in the model only when the industry *as a whole* was unaware how much new supply was in the pipeline. This is an extremely unlikely situation in most chemical product sectors; therefore we rejected this hypothesis. However, exceptions might occur in new industries or obscure sectors where the participants are geographically dispersed and which receive little, if any, media or market research coverage.

Overall, our analysis indicated that construction delays and imperfect information make a production/investment cycle of the type commonly observed in commodity chemicals inherently unstable, particularly on the demand side. Once the system is thrown into disequilibrium, it is difficult (if not impossible) to bring it back into balance.

16.2
Managing Cyclicality for Competitive Advantage – a Worthwhile Endeavor?

Most managers in the chemical industry realize that eliminating cyclicality is desirable, but that it is highly unlikely to happen because the principal drivers are so deeply rooted. It would require, for example, product sectors to consolidate to a point where only a few companies (a maximum of three or four) control global supply and demand – and for those companies to exert tight pricing and investment discipline over long periods, using only legal methods. There are few, if any, instances of this happening.

So if cyclicality is here to stay, the big question for managers in the chemical industry is: can an individual company manage it to gain competitive advantage? To look for possible answers, we used an expanded version of the business dynamics model of a cyclical business to explore how returns might be improved through active cycle management.

The model was structured to simulate a single company with an initial market share of 20 percent of a typical product (again, terephthalic acid was used as the illustrative example). This company competed for market share in the troughs and for economic surplus in the peaks. In this way, the dynamics of competitive responses were captured, especially when the company attempted to expand its market share rapidly through aggressive investment.

We used it to look particularly at the effect of having finance available for investments at the "right" time in the cycle, but also at the impact of forecasting demand and price. We also looked at the related area of cyclical asset trading. Overall, the difficulties of predicting cycles reliably combined with the practical difficulty of obtaining finance during downturns meant that any benefits of cyclicality management were far outweighed by more conventional sources of advantage.

16.2.1
Getting Investment Timing Right

The model was used to simulate and compare two competitive situations in which (a) the single player had some kind of structural advantage, and (b) the player attempted to "manage the cycle" by optimizing his investment timing.

In the first scenario, the company was given the advantage of having either variable costs that were 20 percent lower than the industry average, or fixed costs that were 40 percent lower than the average, while being subject to the same investment constraints as the rest of the players in the industry. Simulation over a period of 60 years showed that a company with this kind of structural advantage would enjoy an average performance lead (measured as cumulative return on invested capital minus the weighted average cost of capital) of around 5 percentage points over competitors (Fig. 16.5).

Structural advantages of the magnitude assumed in the first scenario – arising from feedstock, location or proprietary technology – are reasonably common in the chemical industry. They explain, for example, the dominance of the Middle East, the US Gulf Coast, Western Canada, and certain other locations in the production of basic and intermediate petrochemicals. Being realistic, if a company possesses advantages of this sort, sustaining them is likely to be easier and more

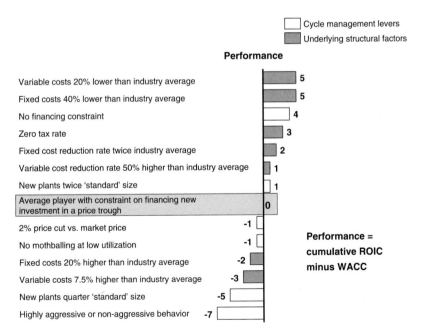

Fig. 16.5 Relative financial performance advantage gained by having structural advantages *versus* pulling cycle management levers
Source: McKinsey

rewarding than the much greater task of attempting always to invest at the "right" time in a cycle.

In the second "cycle management" scenario, the financing constraints were relaxed so that the player could make investments at the "right" time in the cycle (i.e., it could bring new capacity onstream just as prices were about to fly up), while having the same cost structure as the rest of the industry. This strategy proved a sound one, yielding a performance advantage of 4 percentage points over competitors. However, it is extremely difficult to execute such a strategy consistently over time, since it requires an extraordinary level of insight and foresight about both future demand and future supply.

There are, in addition, severe financial and other constraints on getting investment approvals at the "right" time. Optimizing investment timing may mean, for example, having to rethink the relationship between business units and the corporate center so that long term business needs – such as committing funds for building capacity during a trough – are not overridden by short term corporate bookkeeping considerations.

In order to be able to identify the golden moment for investment, business units also need to be able to forecast demand more accurately – a capability that is often underdeveloped in chemical companies. Investing relatively small sums to understand the dynamics of demand for chemical products (such as price elasticity with competing materials) could go a long way toward reducing the volatility of profits in a cyclical environment, since an accurate picture of future demand will help companies to adjust supply to match demand in less major ways: for example, by expanding plants already in operation, de-mothballing others, or buying product from other sources to achieve more level operating rates and more stable pricing.

16.2.2
Price Prediction as a Strategic Tool

Chemical industry managers are only too well aware of the ups and downs of prices and margins in commodity sectors. However, few of them stay long enough within a single product sector to experience more than one major cycle and understand the complex – and often product-specific – factors that contribute to volatile margins. Even more importantly, companies' management information systems rarely contain data of sufficient quality to allow consistent analysis of a business unit's average long term performance during several cycles – over, say, 15–20 years (or even longer).

It is not unusual, therefore, to find chemical companies making decisions on whether to continue ownership or investment in cyclical sectors on the basis of only a few years' historic figures. Yet such short term data inevitably reflect performance over only a part of one cycle, and are potentially highly misleading about the average returns achieved historically (and likely to be achieved in the future) by a particular business unit. If, however, companies were to attempt to predict price – with its implications for margins and returns – over the 15–20 year period mentioned above, this could prove a valuable strategic tool for determining the sectors in which they want to play. Moreover, it appears that such forecasts can be performed with high to reasonable accuracy for a number of important products on the basis of reinvestment economics.

Microeconomic theory maintains that, in the long run, the price of a commodity is going to be equal to the level at which the lowest cost producers can justify incremental investment – known as "reinvestment economics". The argument has it that, unless the average long-run price is equal to that of reinvestment economics, then it will be impossible to attract the investment in new supply to satisfy the incremental growth in demand. At the same time, the long-run price is unlikely to remain above reinvestment economics for long periods because in such a case too much new capacity would be attracted and players would compete (on price) for market share – ultimately driving prices down to the equilibrium "reinvestment" level.

We tested whether it was possible to make long term projections of chemical prices from an understanding of likely future reinvestment economics. Historical prices were compared with actual reinvestment economics for "leader" players in 10 commodity sectors in Western Europe over 20 years from 1980 – a period which contained at least two cycles. The sectors analyzed included major intermediates such as styrene and terephthalic acid, plus polyolefines and polystyrene.

The conclusions (Fig. 16.6) were that, for many products, for example, styrene (Fig. 16.7), polystyrene, acrylonitrile, LLDPE, and PTA, the ability of leader reinvestment economics to predict market prices was remarkably good. A difference between the two of less than USD 50 per tonne on average was observed over the 20 year period, with a similar directional trend. For other products, HPDE and

Long term predictive power of leader re-investment economics	Product	Average 20 year difference between actual price and price predicted by leader reinvestment economics USD/tonne
Good	Styrene	14
	PTA	16
	Acrylonitrile	18
	LLDPE	24
Fair	Polystyrene	56
	LDPE	76
	HDPE	102
	EO/EG	112
Poor	Polypropylene	196
	PET	199

Fig. 16.6 Ability to project future commodity chemical prices from reinvestment economics
Source: McKinsey

USD/tonne

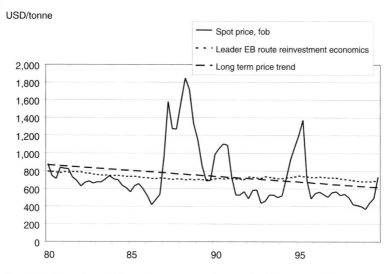

Fig. 16.7 Comparison of long term styrene price trend with styrene reinvestment economics, 1980–99
Source: TECNON, McKinsey analysis

LDPE for example, the predictive power was also reasonably good (USD 50–100 per tonne difference on average).

However, for a third group of products (PET and polypropylene), the predictive power was less good. This can be attributed to a number of possible causes, some of which may be unique to individual products. For example, in the case of polypropylene the observed market price was substantially higher than reinvestment economics would predict. This is probably because of the exceptionally strong historic growth in demand for PP, causing (until the late 1990s) chronic shortages of capacity. This led to high market prices compared to reinvestment economics – as supply only just managed to keep pace with growth in demand.

However, while PET experienced even higher growth rates over a similar period than polypropylene, it was accompanied by a rapid fall in barriers to entry (the technology became widely available and requires relatively little capital) and in reinvestment costs for existing players. The rate at which the capital cost of new PET resin capacity has been declining (over 5 percent per year in real terms between 1982 and 1999, and even faster than that in the mid/late 1980s) is by far the highest of any of the chemical products studied – and this has tended to attract many new entrants.

Our overall conclusion is that, with some exceptions, a 20 year analysis of historic reinvestment economics for a leader producer does give a good indication of the magnitude and direction of future average product prices. It does not, of course, give any indication of the size and timing of future price fly-ups. For that kind of prediction, there is no substitute for detailed microeconomic analysis of historic price setting mechanisms, region by region, and even this is no guarantee of success.

16.2.3
If Managing Cyclicality is not the Best Route to High Returns, What is?

Guided by analysts, and for all the reasons outlined above, investors in chemical industry stocks are wary of companies with highly cyclical earnings. Not only is cyclicality perceived to be difficult to manage, and therefore risky: cyclical chemical companies are also viewed as delivering lower total returns than non-cyclical companies.

To test whether this view reflects actual company performance, we undertook a detailed analysis of the historical performance of the top 100 (by sales) US and non-US chemical companies over the 1987–98 period.

The results were revealing. We found that there is no statistically significant difference between the performance of more cyclical and less cyclical companies either in terms of total returns to shareholders or ratio of market value to invested capital (Fig. 16.8).

At the same time, we determined that top-quartile operational performance is more important for creating shareholder value than is the type, size, or cyclicality (Figs. 16.9 and 16.10) of a chemical company's portfolio.

These findings support the earlier conclusion that ensuring top operational performance (probably through some kind of structural advantage) in a highly cyclical industry is a more rewarding strategic objective than being a second quartile performer in a less cyclical business. We find that the top performers in any

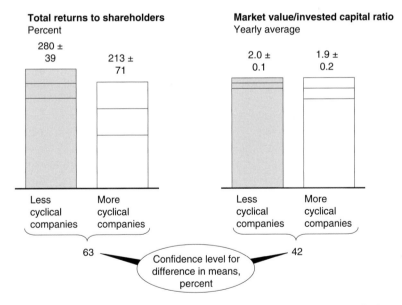

Fig. 16.8 Comparison of total shareholder returns and market/invested capital ratios for less cyclical and more cyclical chemical companies, 1987–98
Source: McKinsey

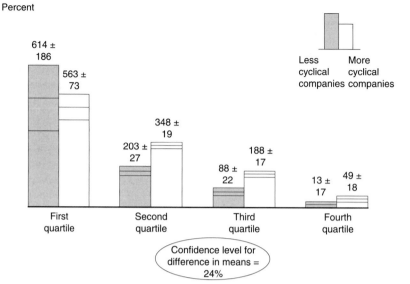

Fig. 16.9 Total shareholder returns in percent (for 100 US and non-US chemical companies) by quartile, for less cyclical and more cyclical companies, 1987–98
Source: McKinsey

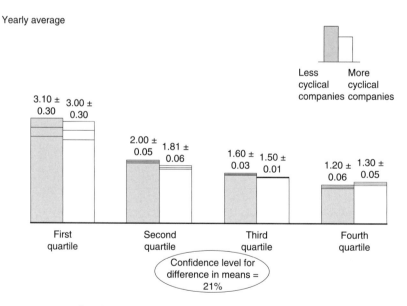

Fig. 16.10 Ratio of market value/invested capital (for 100 US and non-US chemical companies) by quartile, for less cyclical and more cyclical companies, yearly average for 1987–98
Source: McKinsey

chemical business, cyclical or not, are those that exploit leading technologies on a global scale, innovate continuously to maintain technological leadership, and gain long term access to low cost feedstocks. Struggling to manage cyclicality is a worthwhile but – by implication – a secondary endeavor.

16.3
Cyclical Asset Trading

Cyclical asset trading is a further aspect of cycle management which is not directly related to the production and sale of chemicals, but which is considered here because it is practised by some chemical industry players and can, of course, affect the industry. It echoes the investment timing problem: is it possible to create value on a consistent basis by acquiring assets "cheaply" at a low point in the cycle and then reselling them at its peak? Much has been made of the strategies of private companies such as Huntsman, which has reputedly negotiated many acquisitions of chemical assets at a significant discount to their replacement value.

In theory, a cyclical asset trading strategy is simple – "buy low, sell high". In practice, it is likely to be more difficult to achieve for a number of reasons:

- It requires the perceived values of fixed assets to be consistently higher at the top of a cycle than in a trough. While this is superficially and psychologically plausible, a reasonably thorough discounted cash flow (DCF) valuation will show very little difference in the value of a business at either a trough or peak. This is because the bulk of the value lies well beyond the first cycle of two or three years of operations. Besides, we are not aware of any reliable body of data that demonstrates that prices actually paid for chemical assets are systematically higher at the peaks of cycles than in the troughs.
- It requires a reasonably liquid market in chemical assets, that is, one in which there are many buyers and sellers. In reality, an increasing strategic focus by chemical companies on a few sectors is severely limiting the number of trade buyers who might be interested. Besides, antitrust considerations often place limitations on "obvious" acquirers who could extract the maximum synergies (plus the expected future trading value) from a deal. Furthermore, physical constraints at integrated chemical sites pose separation problems, while the commonly encountered need to negotiate feedstock supply agreements between old and new owners complicates asset valuations enormously.
- Most importantly, it requires a seller to be able to predict the peak of the next trough with greater accuracy than any potential acquirer, and to prepare an asset for sale at the "right" time. This is a skill that is similar to the timing of new capacity additions: it has the potential to create great value, but is enormously difficult to execute on a consistent basis.

There is yet another constraint analogous to the difficulty of timing new investments in commodity industries – namely, the fact that the capital for asset trading has to be available in a trough, the worst possible time for most organizations. This effectively rules out conventional publicly-held chemical companies, which are reluctant to have investors observe such unconventional behavior.

In fact, chemical companies which adopt a "trade mispriced assets" strategy are not really competing against chemical industry rivals at all, but against other financial investors. This therefore effectively limits cyclical asset trading to private companies, which – if they are to be successful – require an exceptionally highly developed set of skills for analyzing and predicting cycles, plus a minimum understanding of how to compete in chemical product markets.

In summary, it seems that in the longer term cyclicality probably does not affect total returns to shareholders, though it does of course pose management problems in balancing income against investment as well as in explaining the causes and consequences of cyclical prices to stakeholders. In terms of investing management effort, it is much more worthwhile to concentrate on maintaining or creating advantages in cost, technology, and innovation leadership and feedstock accessibility.

Conclusion

Having said all this, the question still remains: where will the chemical industry go from here? Will it succeed in restarting the growth engine? When and how will e-commerce really transform the industry's ways? When will biotechnology research come up with commercially viable, large-scale products and processes, and once more revolutionize an industry that has been revolutionized many times before?

One thing is certain: for the leaders, there will always be opportunities to create value and to develop and implement new ideas. Chemical industry managers who keep an open mind and are prepared to exploit the opportunities created by an ever changing environment will be most capable of maintaining and managing a pipeline of growth options, not only inside the chemical industry but also at its periphery. Those who have their core businesses in order and run well-functioning organizations with highly developed executional skills will find it easier to exercise such options. Those who combine these two strands will be best placed to create value to all stakeholders, outperforming both rival chemical companies and players in other industries.

Finally, in repositioning themselves to take up new growth options, managers are likely to find new issues appearing on the radar screen, demanding radically different skill sets from the successful models of the past. At present, forward-looking chemical companies may need to focus on ways to attract a larger share of high potential staff in the heightening War for Talent. A more innovative approach to R & D may call for the development of new institutional skills such as venture management, and business building capabilities could well be needed to help establish new growth activities. Managers need to be on the alert to recognize and capitalize on such new developments, both now and in the future.

Given its sheer size and diversity as well as its proven ability to adapt to discontinuities in the past, we firmly believe that the chemical industry will succeed in mastering such challenges. We hope that our book will make some contribution toward this successful outcome.

The Editors

Index